Post-harvest Physiology & Quality Management of Fruits and Vegetables

Post-harvest Physiology & Quality Management of Fruits and Vegetables

Editor

V. D. Shinde

scitus academics

Post-harvest Physiology & Quality Management of Fruits and Vegetables

Edited by **V. D. Shinde**

ISBN: 978-1-68117-089-3

Library of Congress Control Number: 2015935488

SCITUS Academics LLC,
www.scitusacademics.com
Box No. 4766, 616 Corporate Way,
Suite 2, Valley Cottage,
NY 10989

Printed in United States of America on Acid Free Paper ∞

Contents

Preface

Postharvest priorities across the globe have evolved considerably over the past four decades, from being exclusively technical in their outlook, to being more responsive to consumer demand. Consumer-driven trends which have contributed to this shift include rising incomes in urban areas, changing dietary habits, more women in the work-place, reduced time for meal preparation and growing demand for safety, quality and convenience. Other factors such as globalization, urbanization and the need to achieve efficiencies and reduce costs have also contributed significantly to this shift in priorities and continue to re-shape and restructure the fresh produce sector. Growing populations across the globe continue to create demand for fresh produce and processed horticultural products. Meeting these requirements as well as those of export markets necessitates assuring quality and safety in both domestic and export supply chains. Capacities must therefore be developed across the region in order to respond to consumer and market demand and to avert the risk of large numbers of small farmers becoming marginalized. At the same time, Governments in the region must develop a vision for the development of the postharvest sector and facilitate activities within the sector in order to realize that vision.

Editor

Chapter 1

Investigation of Approaches to Preserve Postharvest Quality and Safety in Fresh-cut Fruits and Vegetables

H.S. Hajizadeh[1] and M. Kazemi[2]

[1]Department of Horticultural Sciences, College of Agriculture, University of Nlaragheh, Maragheh, Iran

[2]Young Researchers Club, Karaj Branch, Islamic Azad University, Karaj, Iran

ABSTRACT

Fresh cut fruits and vegetables raise food safety concerns including physiological and pathological limitations since they are often processed in a ripe stage which makes them more susceptible to quality deterioration. In this review, we investigated some advances for the maintenance of fresh-cut fruits and vegetables quality, including synthetic chemicals or plant natural antimicrobials and antioxidants, calcium salts, hot water and anoxia treatments, different type of coating and atmosphere conditions. In this review Phenylalanine Ammonia-lyase (PAL), Polyphenol Oxidase (PPO) and membrane permeability were discussed. Although, increasing in browning enzymes activities is documented in some cases but it seems that the main reason for browning inhibition of the fresh-cut products may be because of the maintenance of compartmentation between enzymes and their substrates, preventing enzymatic reaction.

INTRODUCTION

Fresh-cut quality can be affected by both internal (e.g., morphological, physiological and biochemical defense mechanisms, genotype and processing maturity) and external factors (e.g., storage temperature, humidity, cutting-knife sharpness and chemical treatments) (Rolle and Chism, 1987; Watada and Qi, 1999; Emongor, 2004; Mahdavian *et al.*, 2007; Mba *et al.*, 2007; Safizadeh *et al.*, 2007; Kazemi *et al.*, 2011a-f; Shirzadeh and Kazemi, 2011). Hence, they underlying some modifications such as discoloration, decline in levels of nutrients such as ascorbate (Klein, 1987), development of off-odors, membrane breakdown, tissue softening and microbial contamination (Brecht, 1995; Varoquaux and Wiley, 1994; Ruiz-Cruz *et al.*, 2010) after processing. By recognizing and controlling factors that have a deteriorative effect on quality and use of natural ways to reduce deterioration and decay without use of undesirable chemicals, good quality fresh-cut product with sufficient shelf-life can be attained. So, the primary solutions for resolving these problems and fresh-cut responses to the applied treatments will be addressed in this review.

NATURAL METHODS TO PRESERVE FRESH-CUT PRODUCTS

Temperature, essential oils, modified atmospheres and packaging are more natural prevail approaches for keeping quality of fresh-cut fruits and vegetables.

Temperature

Temperature is one of the most effective factors, controlling freshness and decay of the fresh-cut products and maintains good temperature control (the cool chain) is the firs factor in extending shelf-life of high quality fresh-cut products. Fresh-cut product invariably will be subject to temperature abuse during transportation, distribution and retail (Toivonen *et al.*, 2001).

Dea *et al.* (2010) results suggest that the Heat Water (HW) quarantine treatment applied to whole mangoes does not significantly affect the quality of fresh-cut Kent' mango slices stored at 5°C.

However, keeping them at temperatures between 0-10°C can reduce the chance of bacterial and fungal attack since cool temperatures keep some harmful microorganisms at bay but the cold can also cause injury.

Rivera-Lopez *et al.* (2005) showed that in storing Fresh-cut papaya at 5 and 10°C prevented loss of firmness. Also, Allong *et al.* (2000) demonstrated that a storage temperature of 5°C was more effective than 10°C in delaying the ripening and microbial growth while preserving sensory quality of fresh-cut mango slices.

Essential Oils

Methyl Jasmonate (MJ, $C_{13}H_{20}O_3$) is a naturally occurring substance that is commonly found in plant tissues and is a part of the defense mechanism in plants which makes them more resistant to temperature changes and to attacks by insects, bacteria and fungi (Swenson, 1998; Ismail *et al.*, 2011; Ahmad *et al.*, 2005; Musyimi and Ogur, 2008).

It has been reported that MJ treatment can reduce the development of chilling injury symptoms in zucchini (Wang and Buta, 1994) and mango (Gonzalez-Aguilar *et al.*, 2000). In addition, MJ has been shown to suppress the fungal growth in grapefruit (Droby *et al.*, 1999), reduce decay and maintain the postharvest quality of papayas (Gonzalez-Aguilar *et al.*, 2003) and inhibit the microbial contamination of fresh-cut celery and peppers (Buta and Moline, 1998). In raspberry, it is demonstrated that MJ increases the resistance of tissues against decay by enhancing their antioxidant system and their free radical scavenging capability (Wang *et al.*, 1996; Wang and Lin, 2000).

The exposure of the sliced pineapple to a MJ emulsion decreased microbiological growth by 3 logs after 12 days of storage at 7°C, with no effect on firmness or color of the fruit (Martinez-Ferrer and Harper, 2005). The results of Wang and Buta (2003) showed that kiwifruit slices treated with different concentrations of methyl jasmonate, maintained good quality after 3 weeks at 10°C compared to the controls. Also,Moline *et al.* (1997) declared that the most effective treatment in reducing decay in strawberries and in reducing the growth of *Botrytis cinerea* was MJ.

Packaging and Modified Atmosphere

Packaging fresh-cut fruit products in polymeric films that help in creating a suitable passive modified atmosphere can also be an effective supplement to proper temperature management in maintaining their quality (Martinez-Ferrer *et al.*, 2002; Fonseca *et al.*, 2005).

The use of anti-browning dips such as Nature Seal AS-1 can control the surface browning of apple slices as well as improve the texture to a small degree. However, the matching of an appropriate film to bag the slices is important ensure complete control of browning and to control other deteriorative processes. Oxygen levels can get too low in the wrong bag and this can lead to off-flavor development as a result of anaerobic respiration. The challenge is to select a package system which will have low enough oxygen to assist in quality retention without leading to off-flavor development. Generally a range of 2 to 5% oxygen is recommended for apples.

Modified and controlled atmospheres help maintain quality and extend storage life by inhibiting metabolic activity, decay, browning (Gunes and Lee, 1997) and especially by inhibiting ethylene biosynthesis and action (Kader, 1986; Kader *et al.*, 1989; Mathooko, 1995). The most common atmospheres consist of reduced O_2 and elevated CO_2 levels. Carbon monoxide is also sometimes included for inhibition of browning and microorganism growth (Kader *et al.*, 1989).

The project for evaluating the effect of packaging and modified atmosphere on fresh-cut Salad savoy (*Brassica oleracea* L.) showed that packages with 16.6 and 21.4 Oxygen Transmission Rate (OTR) films attained the desired O_2 (1.4-3.8 kPa) and CO_2 levels (3.6-6.3 kPa) on day 10 and throughout the storage period and maintained freshness with high overall quality scores (Kim *et al.*, 2004).

Odriozola-Serrano *et al.* (2010) investigated bioactive composition of fresh-cut strawberries stored under super atmospheric oxygen, low-oxygen or passive atmospheres. They proposed that 2.5 kPa O_2+7 kPa CO_2 atmospheres are better choice to prevent oxidation of the main antioxidant compounds in fresh-cut strawberries.

In the other project on fresh-cut pineapple, the main effect of reduced (8 kPa or lower) O_2 levels was better retention of the yellow color of the pulp pieces, as reflected in higher final chroma values, whereas elevated (10 kPa) CO_2 levels led to a reduction in browning (higher L values). Modified atmosphere packaging allowed conservation of pulp pieces for over 2 weeks at 5°C or lower without undesirable changes in quality parameters (Marrero and Kader, 2006).

Another technology that is being explored is the use of pre-slicing treatment of whole apples with high oxygen. The high oxygen treatment leads to better firmness retention, lowered browning and reduction of anaerobic volatile accumulation in slices made from treated apples (Toivonen *et al.*, 2001). Donadon and Durigan (2004) compared three types of polymeric films for packaging 'Tommy Atkins' mango slices and found that those packed in the polyethylene terephthalate (PET) clamshell trays had a shelf-life of 14 days at 3°C versus 11 days for the mango cubes in the other packages. Singh *et al.* (2007) concluded that the shelf-life of fresh-cut mangoes could be extended by packaging in PET containers. Chonhenchob *et al.* (2007) reported that extended shelf-life was observed in fresh-cut mangoes packed in PET due to reduced oxygen and elevated carbon dioxide concentrations.

Oms-Oliu *et al.* (2008) investigated the effect of alginate, pectin and gellan-based edible coatings on the shelf-life of fresh-cut 'Piel de Sapo' melon. They concluded all coatings prevent of dehydration and ethylene production and did not improve microbiological stability of fresh-cut melon packaged under passive modified atmosphere. In general, pectin-based coating seemed to best maintain sensory attributes.

The application of the edible coatings retarded the microbiological deterioration of fresh-cut apples. Also, Alginate and gellan edible coatings effectively prolonged the shelf-life of Fuji apple wedges by 2 weeks of storage compared with the control apple with 4 days shelf life (Rojas-Grau *et al.*, 2008).

Our researches on intact apple (Hajizadeh *et al.*, 2006) and mushroom (unpublished data), showed that packaging of intact fruits in poly propylene film along with active modified atmosphere were the most effective treatment in extending the shelf life of fruits, so it seems that similar treatment can be used for apple and mushroom slices. Nevertheless application of these treatments needs to more researches and understanding different physiological mechanism between sliced and intact fruit.

CHEMICAL COMPOUNDS TO PRESERVE FRESH-CUT PRODUCTS

Fresh-cut Products and Color Preservation

One of the limitations of fresh-cut fruit products is that they are often processed in a ripe stage, which makes them more susceptible to quality deterioration once they are cut and packaged (Toivonen and DeEll, 2002). On the other hand, This product is subject to enzymatic browning and must be treated with a browning inhibitor to prevent development of unsightly discoloration (Sapers *et al.*, 2002). Enzymatic browning represents a major challenge in fresh-cut fruits (Son *et al.*, 2001; Sapers *et al.*, 2002). Browning occur when the products of phenyl propanoid metabolism, such as various phenolic and possibly other substrates (e.g., anthocyanin) are oxidized in reactions catalyzed by phenolases such as Polyphenoloxidase (PPO) or peroxidases. Chemical dips (such as ascorbic and citric acid, calcium chloride and other compounds) have been shown to be effective in retarding browning and softening of several types of fruit such as apple (Son *et al.*, 2001; Cocci *et al.*, 2006), pineapple (Gonzalez-Aguilar *et al.*, 2004) and pear (Dong *et al.*, 2000;Arias *et al.*, 2008). Arias *et al.* (2008) has recently shown that the use of modified atmospheres combined with an anti-browning treatment (ascorbic acid+4-hexylresorcinol+$CaCl_2$) could be successfully applied to 'Conference', 'Williams' and 'Abate Fetel' pears. Trindade *et al.* (2003) concluded that the most suitable conditions for quality preservation of fresh-cut 'Tommy Atkins' mango were dipping in a solution of 3.5% (w/w) calcium chloride at 35°C for 20 min and packaging under active modified atmosphere (5% oxygen+5% carbon dioxide). Under these conditions, fresh-cut mango maintained good quality for 5 days at 5°C.

Fresh cut apples turn brown when iron-containing chemicals in the apple react with oxygen in the air. We see this every day when iron objects rust. The chemical reaction is called 'oxidation'. There are several ways to inhibit oxidation. Chemical anti-oxidants can be added to food. Lemon juice, for example, will inhibit the browning of a freshly cut apple because lemons are high in citric acid, an anti-oxidant. Sulphur dioxide, used in the commercial processing of many foods, does the same thing. Removing the air (and thus the oxygen) from food can also inhibit oxidation. This is one reason can food stays fresh so long. Because the enzymes involved in oxidation need water to work, the reaction can be also slowed by de-hydrating (removing water from) food.

Apple cultivars have different capacity for browning, hence, choosing apples that are less prone to discoloration is necessary for fresh cut industry. A study found that 'Arangeh' tend to brown the least. 'Granny Smith' and 'Golden Smoty' apples also showed minimal browning. Golden Delicious apples were in the middle and Red Delicious apples browned the most. In Granny, Smith and Fuji wedges were treated with a pH 2.9 dip containing ascorbic acid, citric acid and sodium hexametaphosphate suppressed browning for at least 3 week at 4°C, whereas formulations without hexametaphosphate failed within 1 week (Pilizota and Sapers, 2004). Results showed that incorporation of the antioxidant to the coating reduced browning compared to the use of the antioxidant alone. Increasing AA and Cys content decreased browning of coated samples. The most effective treatments were WPC-BW-based coatings with 1% AA or 0.5% Cys.

Fresh-cut products and texture preservation: Firmness retention is an important quality parameter in fresh-cut fruits and vegetable products (Agar *et al.*, 1999; Gorny *et al.*, 1999). An important aspect of the fresh-cut industry is texture. Fresh-cut produce must have a reasonable shelf-life which is the time between when produce is cut and when it is consumed. If not ripe enough, it won't taste good. But if the product is cut at too ripe stage, then it will deteriorate even more rapidly.

Lamikanra and Watson (2004) study the effect of calcium dipping and temperature on the quality of Fresh-cut Cantaloupe Melon and showed that Fruit dipped in solution at 4°C had lower respiration and moisture loss rates than treated fruit at ambient temperature. Also, Calcium treatment lowered lipase activity at both temperatures but the effect was more notable in fruit treated at the lower temperature.

Botelho *et al.* (2006) investigated the quality of fresh-cut and intact strawberry during storage at 5°C. Fruits were dipped in sodium isocyanurate solution and packaged in polyethylene terephthalate (PET) trays. Results showed that fresh-cut strawberry had lower SSC and firmness, lighter colour and higher respiratory activity than intact fruit and showed good quality for up to 4 d after cutting compare to intact strawberry which stored for 8 days.

Treatment fresh-cut 'Blanquilla' pears with 1-MCP before cutting and peeling considerably improved their textural properties and color and allows fresh cut 'Blanquilla' pears to be sold up to about 5 d after processing. Treatment with 1-MCP could be a viable alternative to common technologies for extending the shelf-life of 'Blanquilla' pears as a fresh-cut product (Arias *et al.*, 2009).

Ergun *et al.* (2007) studied Physiology of fresh-cut 'Galia' (*Cucumis melo* var. *reticulatus*) from ripe fruit treated with 1-methylcyclopropene and demonstrated that 1-MCP treatment deferred loss of physical deterioration of fresh-cut 'Galia' cubes at 5°C by 2-3 d compared with controls.

Fresh-cut Products and Flavor Preservation

Flavor quality of fruits and vegetables is influenced by genetic, preharvest, harvest and postharvest factors. The longer the time between harvest and eating, the greater the losses of characteristic flavor (taste and aroma) and the development of off-flavors in most fruits and vegetables. Postharvest life based on flavor and nutritional quality is shorter than that based on appearance and textural quality. Thus, it is essential that good flavor quality be emphasized in the future by selecting the best-tasting genotypes and using an integrated crop management system including harvest at the maturity or ripeness stage and use the postharvest handling procedures that will maintain optimal flavor and nutritional quality of fruits and vegetables between harvest and consumption (Kader, 2008). Marrero and Kader (2006) researches on the optimal temperature and modified atmosphere for keeping quality of fresh-cut pineapples showed that the very low respiration rates (CO_2 production of about 0.3 L kg^{-1} sec^{-1}) of pulp pieces at 0, 2.2 and 5°C allowed the use of a MA film such as Mylar®, with a low O_2 transmission rate, without development of permanent off-odors or flavors due to anaerobiosis.

SAFETY

Generally, there is a positive correlation between longer shelf-life of fresh-cut fruits and low aerobic plate count, low total plate count and especially low yeast and mold counts. Thus, it is very important to avoid sources of microbial contamination and to wash the fruits with disinfected water before cutting (Kader, 2008). Fresh-cut products are particularly susceptible to microbial growth owing to the removal of plant protective tissues and the release of cellular fluids from cutting (Heard, 2002), which results in shelf life reduction and food-borne illnesses given that fresh-cut products are marketed as prewashed and ready to eat and not subject to further microbial killing steps, the development and proper application of sanitizing agents to remove microorganisms and control pathogen cross contamination effectively is critical to ensure the quality and safety of fresh-cut produce (Cruz *et al.*, 2006).

There are many reports about using of disinfectants for mango such as 80% ethanol (Thambaramaya, 1997; Plotto *et al.*, 2006), peroxyacetic acid (100 ppm) (Narciso and Plotto, 2005), vanillin solutions (Ngarmsak, 2007) before cutting.

A major consideration in safety has been identified as inoculation of the nutrient-rich flesh of vegetables and fruits with human pathogens from the surface tissues, which are less conducive to growth and development than internal tissues. Safety must be of primary concern in any fresh-cut product.

The potential for human pathogen contamination is indicated by numerous outbreaks of Escherichia coli O157:H7 food poisoning associated with unpasteurized apple cider which is found in the orchard environment and may be detected on fresh apples (Riordan *et al.*, 2001) and can grow in wounds on the apple surface (Gunes and

Hotchkiss, 2002), creating opportunities for preharvest contamination. It has been established that *Clostridium botulinum* can grow on romaine lettuce and shredded cabbage in pouches (Petran *et al.*, 1995). Survival and growth of *L. monocytogenes* on apple slices was demonstrated by Conway *et al.* (2000). *Listeria innocua* was able to survive and grow in a chilled, neutral pH browning inhibitor solution similar to that widely used by fresh-cut apple processors (Karaibrahimoglu *et al.*, 2004). So, use of an acidic formulation probably would have suppressed survival and growth of *L. monocytogenes* in the dip, thereby preventing cross contamination of the fresh-cut product (Pilizota and Sapers, 2004).

Ruiz-Cruz *et al.* (2010) investigated the effect of four sanitizers include Sodium hypochlorite (OCl), Peroxyacetic Acid (PA), Acidified Sodium Chlorite (ASC) and carvacrol on microbiological, sensorial and nutritional quality (total phenols, vitamin C and antioxidant capacity) of fresh-cut jalapeno peppers and stored them at 5°C for 27 days. They demonstrated that all sanitizers (except for carvacrol) maintained microbiological and overall quality of jalapeno peppers and ASC was the most effective sanitizer even though it was used at concentrations lower that those currently approved by the FDA.

DISCUSSION

Important considerations in quality of fresh-cut items include a consistent and fresh appearance, acceptable texture, characteristic flavor and sufficient shelf-life to survive the distribution system. Cutting of produce removes the natural protection of the epidermis and destroys the internal compartmentation that protects enzymes from substrates. The consequences of cutting of vegetables appear to be less severe than the cutting of fruits (Brecht, 1995). Microbial decay is the major source of spoilage of fresh-cut produce, since washing and chlorinated water dips only partially remove the microorganisms intrinsic to produce. All sanitizers were capable of controlling microbial growth without inducing major loss of antioxidant capacity and phytochemicals. Carvacrol was the only sanitizer that reduced sensory acceptability of fresh-cut jalapeno peppers however, carvacrol treated samples retained the highest levels of phytochemical and antioxidant capacity. Calcium application often results in reduced rates of respiration and ethylene production, increased firmness and reduced incidence of physiological disorders and decay (Conway and Sams, 1987; Garcia *et al.*, 1996). The ability of calcium to confer rigidity to the tissue components at low temperatures, possibly through improved covalent cross linking, was indicated by viscosity measurements indicating higher values for pulverized cantaloupe melon with added calcium at 4°C than fruit blended under similar conditions at ambient temperature.

Browning-inhibitor formulations generally contain a reducing agent, such as ascorbic acid and also may contain other browning inhibitors or adjuncts, such as citric acid, a calcium salt, cysteine, polyphosphates, or 4-hexylresorcinol (Sapers, 1993; Sapers *et al.*, 2002). Such formulations may be acidic or neutral in pH. One of the most widely used browning inhibitors is a neutral product containing calcium ascorbate, marketed as Nature Seal TM (Chen *et al.*, 2000).

With temperature being the main factor in fresh-cut pineapple conservation, the main advantage of atmosphere modification was a better retention of color. Reduction of O_2 concentration to 8% or lower improved the final appearance as reflected by higher final chroma values. Increasing CO_2 concentration to 10% had the added advantage of retained Luminosity (L) of the pulp pieces. This could be due to a lower activity of browning enzymes such as polyphenol oxidase. Also, elevated CO_2 atmospheres may have delayed microbial growth.

As mentioned before, modified atmosphere packaging is widely used for fresh-cut vegetables and fruits. Semi permeable plastic films are chosen for MAP so that the film permeability and product respiration can combine to produce a desirable steady state atmosphere within the package (Kader *et al.*, 1989). Undesirable atmospheres can lead to reduced quality due to off-flavor and discoloration (Mateos *et al.*, 1993). Because of the relative perishability of fresh-cut products, the atmosphere in the MAP is often actively established, either by flushing with the desired atmosphere or by pulling a slight vacuum and then injecting a desired gas mixture. Novel approaches such as chemical disinfection (Hong and Gross, 1998), edible coatings (Li and Barth, 1998), natural plant products (Buta *et al.*, 1999; Leepipattanawit*et al.*, 1997), ethylene absorbents (Abe and Watada, 1991), gamma irradiation (Chervin and Boisseau, 1994; Hagenmaier and Baker, 1997), heat shock (Loaiza-Velarde *et al.*, 1997), microbial competition (Breidt and Fleming, 1997), non thermal physical treatments (Hoover,

1997) and pulsed-microwave irradiation (Shin and Pyun, 1997) have been studied as potential alternatives or adjuncts to MAP.

Application of aqueous solutions of calcium salts can help maintain fresh-cut tissue firmness (Izumi and Watada, 1995). Although, calcium chloride has been most commonly used for this purpose, it has recently been shown that calcium lactate is as effective without imparting a bitter flavor as calcium chloride can at higher concentrations (Luna-Guzman *et al.*, 1999). Oxidative browning at the cut surface may be the limiting factor in storage of many fresh-cut vegetables and fruits. Low temperatures employed in the handling and storage of fresh-cut fruits suppress many biological processes; however, tissue softening, deterioration and accumulation of hydrolytic enzymes can continue at relatively high rates (Lamikanra *et al.*, 2000; Karakurt and Huber, 2003; Lamikanra and Watson, 2004). Applications of dilute hypochlorite (Ayhan *et al.*, 1998), calcium salt dips (Luna-Guzam and Barrett, 2000; Saftner *et al.*, 2003), controlled or modified atmospheres (Qi *et al.*, 1999; Bai *et al.*, 2003), edible films and coatings (Hoa *et al.*, 2002; Chen *et al.*, 2009) and irradiation (Palekar *et al.*, 2004) have proven to be of benefit in extending the shelf-life of fresh-cut tissues.

In general, it is clear that 1% $CaCl_2$ is a key compound that can be applied for maintaining firmness and extending shelf life of fresh cut fruits and vegetables cubes regarding of literature review. Mangos must be ripened, at least partially (almost ready-to-eat), before cutting to assure better flavor quality in the fresh-cut products (Kader, 2008). So, it seems the time of harvesting can be one of the factors affecting on fresh-cut fruits flavor. Different methods have been used to prevent deterioration and browning of fresh-cut produce. Traditionally, sulphites have been used to prevent browning, however, their use in processed fruits and vegetables was banned by the Food and Drug Administration in 1986 as a result of adverse reactions developed in certain consumers. An alternative is to inhibit browning and deterioration by using natural products derived from plants (Ahvenainen, 1996). Different effective methods on preserving freshness quality of fruits and vegetables with their shelf life were classified in Table 1

Table 1: Different effective methods on preserving freshness quality of fruits and vegetables

Fresh-cut product	Chemical treatment	Shelf-life (days)	Temperature (°C)	Reference
Cantaloupe melon	Calcium dipping	12	4	Lamikanra and Watson (2004)
Strawberry	Sodium isocyanurate solution	Up to 4	5	Botelho et at. (2006)
Pear	1-MCP	Up to 5	6	Arias et al. (2009)
Melon	1-MCP	3-Feb	5	Ergun a at. (2007)
Apple	Ascorbic acid + citric acid + sodium hexameta phosphate	21	4	Pilizota and Sapers (2004)
Kiwifruit	Methyl jasmonate	21	10	Wang and Buta (2003)
Pineapple	Methyl jasmonate	12	7	Martinez-Ferrer and Harper (2005)

With regards to the data, it can be concluded that Methyl jasmonate may be a practical treatment to ensure the safety and the quality of fresh-cut fruits and vegetables if it is applied with low temperatures which most effective temperature with no chilling damage on fruits and vegetables is 5°C as obtained from literature review. This review has attempted to classify existing knowledge, highlighting effectiveness and differences between approaches to preserve quality of fresh-cut products and it is hoped that such a perspective will lead to use of the most effective method with less limitation on quality and safety of fresh-cut fruits and vegetables. All these approaches and coming technologies may offer improvements in the future. Future research developments will lead to improved scope of the fresh-cut program and hopefully reduce processing costs.

REFERENCES

1. Abe, K. and A.E. Watada, 1991. Ethylene absorbent to maintain quality of lightly processed fruits and vegetables. J. Food Sci., 56: 1589-1592.
2. Agar, I.T., R. Massantini, B. Hess-Pierce and A.A. Kader, 1999. Postharvest CO_2 and ethylene production and quality maintenance of fresh cut kiwifruit slices. J. Food Sci., 64: 433-440.
3. Ahmad, N.R., M.A. Hanif and U. Rashid, 2005. Chemical compositional and intra provenance variation for content of essential oil in*Eucalyptus crebra*. Asian J. Plant Sci., 4: 519-523.
4. Ahvenainen, R., 1996. New approaches in improving the shelf life of minimally processed fruit and vegetables. Trends Food Sci. Technol., 7: 179-187.
5. Allong, R., L.D. Wickham and M. Mohammed, 2000. The effect of cultivar, fruit ripeness, storage temperature and duration on quality of fresh-cut mango. Acta Hortic., 509: 487-494.
6. Arias, E., J. Gonzalez, P. Lopez-Buesa and R. Oria, 2008. Optimization of processing of fresh-cut pears. J. Sci. Food Agric., 88: 1755-1763.
7. Arias, E., P. Lopez-Buesa and R. Oria, 2009. Extension of fresh-cut Blanquilla pear (*Pyrus communis* L.) shelf-life by 1-MCP treatment after harvest. Postharvest Biol. Technol., 54: 53-58.
8. Ayhan, Z., G.W. Chism and E.R. Richter, 1998. The shelf life of minimally processed fresh cut melons. J. Food Qual., 21: 29-40.
9. Bai, J., R.A. Saftner and A.E. Watada, 2003. Characteristics of fresh-cut honeydew (*Cucumis xmelo* L.) available to processors in winter and summer and its quality maintenance by modified atmosphere packaging. Postharvest Biol. Technol., 28: 349-359.

10. Botelho, M.C., S.C. Leme, E.E. Nunes, M.B.V. Boas, E.V.B.V. Boas and A.B. Chitarra, 2006. Quality of fresh-cut strawberry during storage. Acta Hortic., 768: 307-309.
11. Brecht, J.K., 1995. Physiology of lightly processed fruits and vegetables. HortScience, 30: 18-22.
12. Breidt, F. and P.H. Fleming, 1997. Using lactic acid bacteria to improve the safety of minimally processed fruits and vegetables. Food Technol., 51: 44-51.
13. Buta, J.G. and H.E. Moline, 1998. Methyl jasmon at extends shelf life and reduces microbial contamination of fresh-cut celery and pepper. J. Agric. Food Chem., 46: 1253-1256.
14. Buta, J.G., H.E. Moline, D.W. Spaulding and C.Y. Wang, 1999. Extending storage life of fresh-cut apples using natural products and their derivatives. J. Agric. Food Chem., 47: 1-6.
15. Chen, C., X. Dong and L. Flaherty, 2000. Nature seal products extend shelf life of freshcut fruits. Proceedings of the Institute of Food Technologists Annual Meeting, June 10-14, 2000, Chicago IL., USA., pp: 130-.
16. Chen, J.Y., L.H. He, Y.M. Jiang, J.F. Kuang and C.B. Lu *et al.*, 2009. Expression of PAL and HSPs in fresh-cut banana fruit. Environ. Exp. Bot., 66: 31-37.
17. Chervin, C. and P. Boisseau, 1994. Quality maintenance of ready-to-eat shredded carrots by γ irradiation. J. Food Sci., 59: 359-361
18. Chonhenchob, V., Y. Chantarasomboon and S.P. Singh, 2007. Quality changes of treated fresh-cut tropical fruits in rigid modified atmosphere packaging containers. Packag. Technol. Sci., 20: 27-37.
19. Cocci, E., P. Rocculi, S. Romani and M.D. Rosa, 2006. Changes in nutritional properties of minimally processed apples during storage. Postharvest Biol. Technol., 39: 265-271
20. Conway, W.S. and C.E. Sams, 1987. The effects of postharvest infiltration of calcium, magnesium, or strontium on decay, firmness, respiration and ethylene production in apples. J. Am. Soc. Hortic. Sci., 12: 300-303.
21. Conway, W.S., B. Leverentz and R.A. Saftner, 2000. Survival and growth of *Listeria monocytogenes* on fresh-cut apple slices and its interaction with *Glomerella cingulata* and *Penicillium expansum*. Plant Dis., 84: 177-181.

22. Cruz, S,R., Y. Luo, R.J. Gonzalez, Y. Tao and G.A. Gonzalez, 2006. Acidified sodium chlorite as an alternative to chlorine to control microbial growth on shredded carrots while maintaining quality. J. Sci. Food Agric., 86: 1887-1893

23. Dea, S., J.K. Brecht, M.C.N. Nunes and E.A. Baldwin, 2010. Quality of fresh-cut kent mango slices prepared from hot water or non-hot water-treated fruit. Postharvest Biol. Technol., 56: 171-180

24. Donadon, J.R. and J.F. Durigan, 2004. Production and preservation of fresh-cut Tommy Atkins mango chunks. Acta Hortic., 645: 257-260.

25. Dong, X., R. Wrolstad and D. Sugar, 2000. Extending shelf life of fresh cut pears. J. Food Sci., 65: 181-186.

26. Droby, S., R. Porat, L. Cohen, B. Weiss, B. Shapiro, S. Philosoph-Hasas and S. Meir, 1999. Suppressing green mold decay in grapefruit with postharvest jasmonate application. J. Am. Soc. Hortic. Sci., 124: 184-188.

27. Emongor, V.E., 2004. Effects of gibberellic acid on postharvest quality and vaselife life of gerbera cut flowers (*Gerbera jamesonii*). J. Agro., 3: 191-195.

28. Ergun, M., J. Jeong, D.J. Huber and D.G. Cantliffe, 2007. Physiology of fresh-cut Galia (*Cucumis melo* var. *reticulatus*) from ripe fruit treated with 1-methylcyclopropene. Postharvest Biol. Technol., 44: 286-292.

29. Fonseca, S.C., F.A.R. Oliveira, J.K. Brecht and K.V. Chau, 2005. Influence of low oxygen and high carbon dioxide on shredded Galega kale quality for development of modified atmosphere packages. Postharvest Biol. Technol., 35: 279-292.

30. Garcia, J.M., S. Herrera and A. Morilla, 1996. Effects of postharvest dips in calcium chloride on strawberry. J. Agric. Food Chem., 44: 30-33.

31. Gonzalez-Aguilar, G.A., J. Fortiz, R. Cruz, R. Baez and Y. Wang, 2000. Methyl jasmonate reduces chilling injury and maintains postharvest quality of mango fruit. J. Agric. Food Chem., 48: 515-519.

32. Gonzalez-Aguilar, G.A., J.G. Buta and C.Y. Wang, 2003. Methyl jasmonate and modified atmosphere packaging (MAP) reduce

decay and maintain postharvest quality of papaya Sunrise. Posthar. Biol. Technol., 28: 361-370.

33. Gonzalez-Aguilar, G.A., S. Ruiz-Cruz, R. Cruz-Valenzuela, A. Rodriguez-Felix and C.Y. Wang, 2004. Physiological and quality changes of fresh-cut pineapple treated with antibrowning agents. Lebensmittel-Wissenschaft und-Technologie, 37: 369-376.
34. Gorny, J.R., B. Hess-Pierce and A.A. Kader, 1999. Quality changes in fresh-cut peach and nectarine slices as affected by cultivar, storage atmosphere and chemical treatments. J. Food Sci., 64: 429-432.
35. Gunes, G. and C.Y. Lee, 1997. Color of minimally processed potatoes as affected by modified atmosphere packaging and antibrowning agents. J. Food Sci., 62: 572-575.
36. Gunes, G.G. and J.H. Hotchkiss, 2002. Growth and survival of *Escherichia coli* O157:H7 on fresh-cut apples in modified atmospheres at abusive temperatures. J. Food Prot., 65: 1641-1645.
37. Hagenmaier, R.D. and R.A. Baker, 1997. Low-dose irradiation of cut iceberg lettuce in modified atmosphere packaging. J. Agric. Food Chem., 45: 2864-2868.
38. Hajizadeh, H., Y. Mostofi, A. Talaei and M. Musavi, 2006. Modified Atmosphere Packaging (MAP) effects on quality maintenance and storage life extension of local Iranian apple Golab Kohanz. Acta Hortic., 768: 111-116.
39. Heard, G.M., 2002. Microbiology of Fresh-Cut Produce. In: Fresh-Cut Fruits and Vegetables: Science, Technology and Market, Lamikanra, O. (Ed.). CRC Press, Boca Raton, USA., ISBN-10: 1587160307, pp: 187-248.
40. Hoa, T.T., M.N. Ducamp, M. Lebrun and E.A. Baldwin, 2002. Effect of different coating treatments on the quality of mango fruit. J. Food Qual., 25: 471-486.
41. Hong, J.H. and K.C. Gross, 1998. Surface sterilization of whole tomato fruit with sodium hypochlorite influences subsequent postharvest behavior of fresh-cut slices. Postharv. Biol. Technol., 13: 51-58.
42. Hoover, D.G., 1997. Minimally processed fruits and vegetables: Reducing microbial load by nonthermal physical treatments. Food Technol., 51: 66-71.

43. Ismail, A., H. Lamia, H. Mohsen and J. Bassem, 2011. Chemical composition of *Juniperus oxycedrus* L. subsp *Macrocarpa* essential oil and study of their herbicidal effects on germination and seedling growth of weeds. Asian J. Applied Sci., 4: 771-779.

44. Izumi, H. and A.E. Watada, 1995. Calcium treatment to maintain quality of zucchini squash slices. J. Food Sci., 60: 789-793.

45. Kader, A.A., 1986. Biochemical and physiological basis for effects of controlled and modified atmospheres on fruits and vegetables. Food Technol., 40: 99-104.

46. Kader, A.A., 2008. Flavor quality of fruits and vegetables. J. Sci. Food Agric., 88: 1863-1868.

47. Kader, A.A., D. Zagory, E.L. Kerbel and C.Y. Wang, 1989. Modified atmosphere packaging of fruits and vegetables. Crit. Rev. Food Sci. Nutr., 28: 1-30.

48. Karaibrahimoglu, Y., X. Fan, G.M. Sapers and K. Sokorai, 2004. Effect of pH on the survival of *Listeria innocua* in calcium ascorbate solutions and on quality of fresh-cut apples. J. Food Prot., 67: 751-757.

49. Karakurt, Y. and D.J. Huber, 2003. Activities of several membrane and cell-wall hydrolases, ethylene biosynthetic enzymes and cell wall polyuronide degradation during low-temperature storage of intact and fresh-cut papaya (*Carica papaya*) fruit. Postharvest Biol. Technol., 28: 219-229.

50. Kazemi, M., E. Hadavi and J. Hekmati, 2011. Role of salicylic acid in decreases of membrane senescence in cut carnation flowers. Am. J. Plant Physiol., 6: 106-112.

51. Kazemi, M., M. Aran and S. Zamani, 2011. Effect of salicylic acid treatments on quality characteristics of apple fruits during storage. Am. J. Plant Physiol., 6: 113-119.

52. Kazemi, M., M. Aran and S. Zamani, 2011. Effect of calcium chloride and salicylic acid treatments on quality characteristics of kiwifruit (*Actinidia deliciosa* cv. Hayward) during storage. Am. J. Plant Physiol., 6: 183-189.

53. Kazemi, M., M. Aran and S. Zamani, 2011. Extending the vase life of lisianthus (*Eustoma grandiflorum Mariachii*. cv. blue) with different preservatives. Am. J. Plant Physiol., 6: 167-175.

54. Kazemi, M., S. Zamani and M. Aran, 2011. Interaction between glutamin and different chemicals on extending the vase life of cut flowers of Prato lily. Am. J. Plant Physiol., 6: 120-125.

55. Kazemi, M., S. Zamani and M. Aran, 2011. Effect of some treatment chemicals on keeping quality and vase-life of cut flowers. Am. J. Plant Physiol., 6: 99-105.

56. Kim, J.G., Y. Luo and K.C. Gross, 2004. Effect of package film on the quality of fresh-cut salad savoy. Postharvest Biol. Technol., 32: 99-107.

57. Klein, B.P., 1987. Nutritional consequences of minimal processing of fruits and vegetables. J. Food Qual., 10: 179-193.

58. Lamikanra, O. and M.A. Watson, 2004. Effect of calcium treatment temperature on fresh-cut cantaloupe melon during storage. J. Food Sci., 69: C468-C472.

59. Lamikanra, O., J.C. Chen, D. Banks and P.A. Hunter, 2000. Biochemical and microbial changes during the storage of minimally processed cantaloupe. J. Agric. Food Chem., 48: 5955-5961.

60. Leepipattanawit, R., R.M. Beaudry and R.J. Hernandez, 1997. Control of decay in modified-atmosphere packages of sliced apples using 2-nonanone vapor. J. Food Sci., 62: 1043-1047.

61. Li, P. and M.M. Barth, 1998. Impact of edible coatings on nutritional and physiological changes in lightly processed carrots. Postharvest Biol. Technol., 14: 51-60.

62. Loaiza-Velarde, J.G., F.A. Tomas-Barbera and M.E. Saltveit, 1997. Effect of intensity and duration of heat-shock treatments on wound-induced phenolic metabolism in iceberg lettuce. J. Am. Soc. Hortic. Sci., 122: 873-877.

63. Luna-Guzam, I. and D.M. Barrett, 2000. Comparison of calcium chloride and calcium lactate effectiveness in maintaining shelf stability and quality of fresh-cut cantaloupes. Postharvest Biol. Technol., 19: 61-72.

64. Luna-Guzman, I., M. Cantwell and D.M. Barrett, 1999. Fresh-cut cantaloupe: Effect of $CaCl_2$ dips and heat treatments on firmness and metabolic activity. Postsharvest Biol. Technol., 17: 201-213.

65. Mahdavian, K., K.M. Kalantari and M. Ghorbanli, 2007. The effect of different concentrations of salicylic acid on protective

enzyme activities of pepper (*Capsicum annuum* L.) plants. Pak. J. Biol. Sci., 10: 3162-3165.

66. Marrero, A. and A.A. Kader, 2006. Optimal temperature and modified atmosphere for keeping quality of fresh-cut pineapples. Postharvest Biol. Technol., 39: 163-168.

67. Martinez-Ferrer, M. and C. Harper, 2005. Reduction in microbial growth and improvement of storage quality in fresh-cut pineapple after methyl jasmonate treatment. J. Food Qual., 28: 3-12.

68. Martinez-Ferrer, M., C. Harper, F. Perez-Munoz and M. Chaparro, 2002. Modified atmosphere packaging of minimally processed mango and pineapple fruits. J. Food Sci., 67: 3365-3371.

69. Mateos, M., D. Ke., M. Cantwell and A.A. Kader, 1993. Phenolic metabolism and ethanolic fermentation of intact and cut lettuce exposed to CO_2-enriched atmospheres. Postharvest Biol. Technol., 3: 225-233.

70. Mathooko, F.M., 1995. Regulation of ethylene biosynthesis in higher plants by carbon dioxide. Postharvest Biol. Tech., 7: 1-26.

71. Mba, F.O., X. Zhi-Ting and Q. Hai-Jie, 2007. Salicylic acid alleviates the cadmium toxicity in Chinese cabbages (*Brassica chinensis*). Pak. J. Biol. Sci., 10: 3065-3071.

72. Moline, H., J. Buta and R. Saftner, 1997. Comparison of three volatile natural products for the reduction of postharvest decay in strawberries. Adv. Strawberry Res., 16: 43-48.

73. Musyimi, D.M. and J.A. Ogur, 2008. Comparative assessment of antifungal activity of extracts from *Eucalyptus globules* and *Eucalyptus citriodora*. Res. J. Phytochem., 2: 35-43.

74. Narciso, J. and A. Plotto, 2005. A comparison of sanitation systems for fresh-cut mango. HortTechnology, 15: 837-842.

75. Ngarmsak, M., 2007. Antifungal activity of vanillin on fresh-cut tropical fruits. Acta Hortic., 746: 409-415.

76. Odriozola-Serrano, I., R. Soliva-Fortuny and O. Martin-Belloso, 2010. Changes in bioactive composition of fresh-cut strawberries stored under superatmospheric oxygen, low-oxygen or passive atmospheres. J. Food Comp. Anal., 23: 37-43.

77. Oms-Oliu, G., R. Soliva-Fortuny and O. Martin-Belloso, 2008. Using polysaccharide-based edible coatings to enhance quality and antioxidant properties of fresh-cut melon. LWT-Food Sci. Technol., 41: 1862-1870.

78. Palekar, M.P., E. Cabrera-Diaz, A. Kalbasi-Ashtari, J.E. Maxim, R.K. Miller, L. Cisneros-Zevallos and A. Castillo, 2004. Effect of electron beam irradiation on the bacterial load and sensorial quality of sliced cantaloupe. J. Food Sci., 69: M267-M273.

79. Petran, R.L., W.H. Sperber and A.B. Davis, 1995. *Clostridium botulinum* toxin formation in romaine lettuce and shredded cabbage: Effect of storage and packaging conditions. J. Food Prot., 58: 624-627.

80. Pilizota, V. and G.M. Sapers, 2004. Novel browning inhibitor formulation for fresh-cut apples. J. Food Sci., 69: SNQ140-SNQ143.

81. Plotto, A., J. Bai, J.A. Narciso, J.K. Brecht and E.A. Baldwin, 2006. Ethanol vapor prior to processing extends fresh-cut mango storage by decreasing spoilage, but does not always delay ripening. Postharvest Biol. Technol., 39: 134-145.

82. Qi, L., T. Wu and A.E. Watada, 1999. Quality changes of fresh-cut honeydew melons during controlled atmosphere storage. J. Food Qual., 22: 513-521.

83. Riordan, D.C.R., G.M. Sapers, T.R. Hankinson, M. Magee, A.M. Mattrazzo and B.A. Annous, 2001. A study of U.S. orchards to identify potential sources of *Escherichia coli* O157:H7. J. Food Prot., 64: 1320-1327.

84. Rivera-Lopez, J., F.A. Vazquez-Ortiz, J.F. Ayala-Zavala, R.R. Sotelo-Mundo and G.A. Gonzalez-Aguilar, 2005. Cutting shape and storage temperature affect overall quality of fresh-cut papaya cv. Maradol. J. Food Sci., 70: S482-S489.

85. Rojas-Grau, M.A., M.S. Tapia and O. Martin-Belloso, 2008. Using polysaccharide-based edible coatings to maintain quality of fresh-cut Fuji apples. LWT-Food Sci. Technol., 41: 139-147.

86. Rolle, R.S. and G.W. Chism, 1987. Physiological consequences of minimally processed fruits and vegetables. J. Food Qual., 10: 157-177.

87. Ruiz-Cruz, S., E. Alvarez-Parrilla, L.A. de la Rosa, A.I. Martinez-Gonzalez, J.D.J. Ornelas-Paz, A.M. Mendoza-Wilson and G.A. Gonzalez-Aguilar, 2010. Effect of different sanitizers on microbial, sensory and nutritional quality of fresh-cut jalapeno peppers. Am. J. Agric. Biol. Sci., 5: 331-341.

88. Safizadeh, M.R., M. Rahemi, E. Tafazoli and Y. Emam, 2007. Influence of postharvest vacuum infiltration with calcium on chilling injury, firmness and quality of Lisbon lemon fruit. Am. J. Food Technol., 2: 388-396.

89. Saftner, R.A., J. Bai, J.A. Abbott and Y.S. Lee, 2003. Sanitary dips with calcium propionate, calcium chloride, or a calcium amino acid chelate maintain quality and shelf stability of fresh cut honeydew chunks. Postharvest Biol. Technol., 29: 257-269.

90. Sapers, G.M., 1993. Browning of foods: Control by sulfites as antioxidants and other means. Food Technol. Chicago, 47: 75-84.

91. Sapers, G.M., K.B. Hicks and R.L. Miller, 2002. Antibrowning Agents. In: Food Additives, Brane, A.L., P.M. Davidson, S. Salminen and J.H. Thorngate (Eds.). 2nd Edn. Marcel Dekker, New York, USA., pp: 543-561.

92. Shin, J.K. and Y.R. Pyun, 1997. Inactivation of *Lactobacillus plantarum* by pulsed-microwave irradiation. J. Food Sci., 62: 163-166.

93. Shirzadeh, E. and M. Kazemi, 2011. Effect of malic acid and calcium treatments on quality characteristics of apple fruits during storage. Am. J. Plant Physiol., 6: 176-182.

94. Singh, S., V. Chonhenchob, Y. Chantanarasomboom and J. Singh, 2007. Testing and evaluation of quality changes of treated fresh-cut tropical fruits packaged in thermoformed plastic containers. J. Testing Eval., Vol. 35, No. 5. 10.1520/JTE100166

95. Son, S.M., K.D. Moon and C.Y. Lee, 2001. Inhibitory effects of various antibrowning agents on apple slices. Food Chem., 73: 23-30.

96. Swenson, D., 1998. Compound shows promise in extending freshness. The Packer-Inside, Fresh-Processing 8, 9B.

97. Thambaramaya, V.G., 1997. Minimal processing of fruit (apple, custard apple, mandarin, mango, pawpaw). Dissertation Abstr. Int., 58: 1037-1037.

98. Toivonen, P.M.A. and J.R. DeEll, 2002. Physiology of Fresh-Cut Fruits and Vegetables. In: Physiology of Fresh-Cut Fruits and Vegetables: Science, Technology and Market, Lamikanra, O. (Ed.). CRC Press, Boca Raton, FL., pp: 91-123.

99. Toivonen, P.M.A., P. Delaquis, M. Cliff, A.L. Moyls and T. Beveridge, 2001. Factors involved in developing apple slices. Proceedings of the Washington Tree Fruit Postharvest Conference, March 13-14, 2001, Wenatchee, WA., USA., pp: 1-5.
100. Trindade, P., M.L. Beiraro-da-Costa, M. Moldao-Martins, M. Abreu, E.M. Guncalves and S. Beiraro-da-Costa, 2003. The effect of heat treatments and calcium chloride applications on quality of fresh-cut mango. Acta Hortic., 599: 603-609.
101. Varoquaux, P. and R.C. Wiley, 1994. Biological and Biochemical Changes in Minimally Processed Refrigerated Fruits and Vegetables. In: Minimally Processed Refrigerated Fruits and Vegetables, Wiley, R.C. (Ed.). Chapman and Hall, New York, USA., pp: 226-267.
102. Wang, C.Y. and J.G. Buta, 1994. Methyl jasmonate reduces chilling injury in Cucurbita pepo through its regulation of abscisic acid and polyamine levels. Environ. Exp. Bot., 34: 427-432.
103. Wang, C.Y. and J.G. Buta, 2003. Maintaining quality of fresh-cut kiwifruit with volatile compounds. Postharvest Biol. Technol., 28: 181-186.
104. Wang, H., G. Cao and R.L. Prior, 1996. Total antioxidant capacity of fruits. J. Agric. Food Chem., 44: 701-705.
105. Wang, S.Y. and H.S. Lin, 2000. Antioxidant activity in fruits and leaves of blackberry, raspberry and strawberry varies with cultivar and developmental stage. J. Agric. Food Chem., 48: 140-146.
106. Watada, A.E. and L. Qi, 1999. Quality of fresh-cut produce. Postharvest Biol. Technol., 15: 201-205.

Chapter 2

Fundamental Aspects of Postharvest Heat Treatments

Susan Lurie[1] and Romina Pedreschi[2]

[1]Department of Posthavest Science, The Volcani Center, ARO, Bet Dagan 50250, Israel

[2]Pontificia Universidad Católica de Valparaíso, Escuela de Agronomía, La Palma, Quillota, Chile

ABSTRACT

Heat treatments have been investigated for use in many aspects of postharvest storage. They have been developed for insect control, prevention of fungal development and prevention of postharvest storage disorders including chilling injury. The treatment times and temperature range vary widely, from days at 35 °C to 39 °C in hot air, to up to 63 °C for less than a minute in hot water. Much of the research has been performed to develop solutions to a particular problem, and less investigation has been conducted on the responses of the commodity to the treatment. However, since the turn of the century, a number of groups have been active in examining the molecular responses and changes that occur in commodities during and after the heat treatment. This review examines the changes at the level of transcriptome, proteome and metabolome that occur in response to the different heat treatments.

INTRODUCTION

Postharvest heat treatments of fruit are used for insect disinfestation, disease control, to modify fruit responses to cold stress and maintain fruit quality during storage[1, 2, 3, 4, 5] The temperature and exposure duration for a particular crop for a particular purpose are usually determined empirically. Conditioning treatments at temperatures from 30 °C to 40 °C in hot air (HAT) for times ranging from hours to days have been developed to affect commodity quality and storability[1] Higher temperatures are normally used for either insect or microorganism control. Some treatments which have commercial approval for insect control involve either vapor heat or hot water immersion of the fruit until the core temperature reaches 47 °C, which takes a number of hours unless radiofrequency is involved[5] Other treatments, involving hot water dips (HWT), are for a few minutes at temperatures of 50–56 °C to control fungal pathogens. In addition, there is a short (10–25 s) treatment of hot water rinsing and brushing (HWB) which may involve temperatures of up to 63 °C[4]

Much of the published research describes the phenological and physiological responses of a fruit or vegetable to heat treatment. In some studies, biochemical data were gathered on heat shock proteins (HSPs), and other stress proteins such as pathogen resistant (PR), antioxidants, *etc.* However, the greatest number of studies have been concerned with developing an effective treatment against a problem, be it insect pest, pathogen attack or storage disorder.

This review will focus on the molecular responses of fruit and vegetables to postharvest high temperature stress. It will summarize what has been published in the twenty-first century, and emphasize the 'omics' studies; metabolomics, proteomics and transcriptomics. It will not deal with research on the effect of high temperature on minimally processed produce, which is of interest in itself. Nor will it deal with heat damage, which is also of importance[6] The body of knowledge that is being gathered in these studies will help in developing more accurate treatments with less possibility of unwanted side effects to the commodity.

HEAT STRESS RESPONSE

Temperature is one of the most important environmental factors that regulate plant growth and development. Plants have evolved signaling pathways to sense changes in ambient temperature and adjust their metabolism and cell function to prevent heat-related damage. Many features of the heat stress response pathway are conserved among both prokaryotic and eukaryotic organisms.[7]Because plants are sessile organisms that cannot escape stress conditions, they invest valuable resources in modifying their metabolism to prevent damage caused by heat, in a process referred to as acclimation, or acquired thermotolerance.[7, 8] The accumulation of HSPs, under the control of heat stress transcription factors (HSFs), plays a central role in the heat stress response and in acquired thermotolerance in plants and other organisms.

Heat stress affects the stability of various proteins, membranes, RNA species and cytoskeleton structures, and alters the efficiency of enzymatic reactions in the cell, causing metabolic imbalance.[9,10,11] Disrupting the steady-state flux of metabolites can cause the accumulation of toxic compounds, such as reactive oxygen species (ROS). In fact, ROS is generally a component of heat stress and increases in antioxidant processes are part of the heat stress response.[8, 12] As a response to heat stress, plants reprogram their transcriptone, proteome, metabolome and lipidome (Figure 1). These changes establish a new steady-state balance of metabolic processes that can enable the organism to function at the new temperature. This new balance may also be helpful in responding to other stresses, such as the low temperatures of storage.

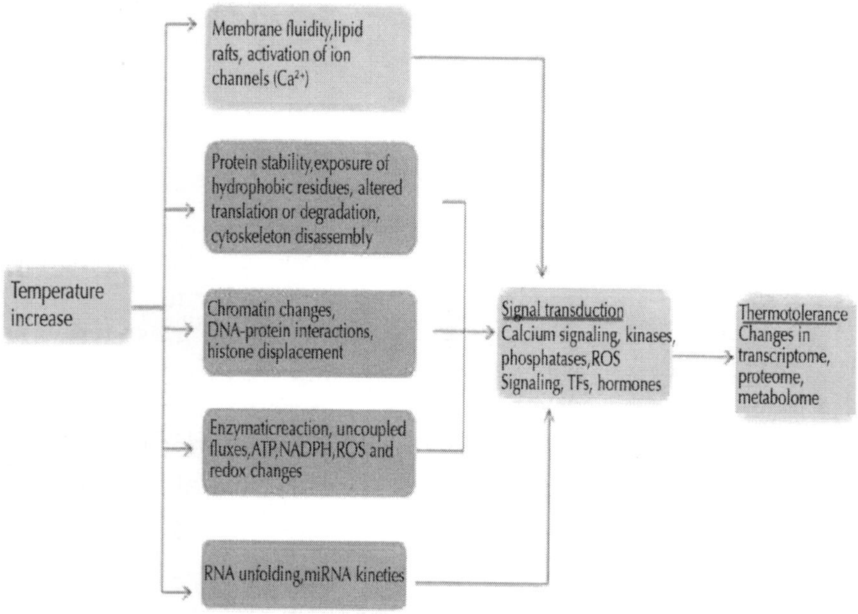

Figure 1: A schematic model for temperature sensing in plants. Changes in temperature can alter membrane properties and activate a calcium channel. The inward flux of calcium that follows will activate signal transduction events and alter plant metabolism to achieve thermotolerance. Additional temperature-induced changes include alterations in protein stability and exposure of hydrophobic residues of proteins that may trigger ubiquitin in the cytosol and the ER, histone displacement which allows HSFs to bind to DNA, accumulation of ROS and alteration in cellular energy levels, and unfolding of RNA species that could affect miRNA. These pathways will also trigger different signal transduction events and lead to thermotolerance (adapted from Mittler *et al*[8] 2012).

A high temperature will affect most macromolecules and will cause changes such as increased membrane fluidity, partial melting of DNA and RNA strands, protein subunit disassociation and exposure of hydrophobic cores. The specific change which might be a sensor of high temperature stress is unknown, and may be different in different plants. The classical model for heat stress activation involves HSFs. Heat-stressed unfolded proteins in the cell cause molecular chaperones to be released from their constitutive inhibitory association with HSF monomers and to bind the unfolded proteins, while the free HSF subunits trimerize, undergo phosphorylation and bind to heat shock response promoters in the genome to activate heat stress responses.[13] However, mild heat stresses may not lead to protein unfolding and yet the plant shows a stress response. Other models which appear to be dominant in some situations include; a specific calcium channel opening in the plasma membrane due to heat-induced changes in membrane fluidity[14] lipid signaling due to membrane fluidity changes[15,16] and ROS signaling, particularly the respiratory burst mediated by a plasma membrane NADP oxidase[12] It is noteworthy that these models all include some action at the plasma membrane, which is also the suggested site of response to cold stress. Recently, there has been some evidence that smRNAs may also be involved in responses to heat stress, although so far only as response to high temperature and not involved in acquired thermotolerance[17,18]

Attempts to increase thermotolerance by overexpression of a single gene have limited impact because of the genetic complexity of the heat stress response[19] Data from genome wide expression profiles of *Arabidopsis* plants exposed to heat stress, from different plant species experiencing heat stress, from different tissue types, developmental stages and different duration of the stress, give some commonality of patterns of transcript accumulation, with between 2% and 5% of the genome being affected[20, 21] Therefore, it would be unlikely that one gene could cause more than a qualitative change in the response of an organism to heat stress.

HEAT TREATMENT OF POSTHARVEST COMMODITIES

Harvested commodities are invariably exposed to low temperature in order to retard product respiration and delay ripening and senescence. Many commodities will develop chilling injury if the temperature is too low or if the cold conditions are maintained for too long. Heat treatments have been found many cases to delay or prevent the development of chilling injury[1] This has been shown in many commodities to be associated with the prolonged presence of HSPs in the tissue and the protective effect they exert[2, 22, 23, 24, 25] HSPs increase during a heat stress and generally disappear rapidly when a plant is returned to ambient temperature. Sabahat *et al*[26] were the first to show that if a commodity was placed at 2 °C rather than 20 °C following a heat stress the HSPs were not metabolized. Thus, the thermotolerance induced in a heat stress can give protection against cold stress.

Postharvest heat treatments also alter the normal program of protein synthesis and cellular metabolism during heat stress. When heat stress is imposed, there is a rapid disassociation of polyribosomes and protein synthesis stops briefly and then resumes with a new set of proteins including the HSPs[27] The consequence of this switch is that normal ripening processes are inhibited, and, if the commodity is then placed in low temperature, the inhibition persists for some time. When rewarmed after storage ripening resumes. Therefore, postharvest heat treatments can modulate the rate of ripening of commodities, while, in addition, preventing postharvest storage disorders.

Studies of changes in transcriptome, proteome and metabolome of a number of fruits due to a postharvest heat treatment have been published. The commodities examined include citrus, mango, peach, potato and tomato (Table 1).

Table 1: Molecular studies of postharvest heat treatments

Commodity	Heat treatment	Methods	Main findings	Reference
Citrus				
*Citrus unshiu*Marc. cv. Kamei Satsuma	HWT 2 min 52 °C	Proteomic 2-DE and MALDI-TOF/TOF Metabolomic LS-MS, HPLC-MS	Stress/defense proteins upregulated (HSPs, PRs); sugars and fatty acids increased; ornathine increased; flavonoids increased and H_2O_2decreased in HT fruit	Yun *et al.*, 2013[32]

Citrus sinensis cv. Valencia	HAT 48 h 37 °C	Proteomic 2-DE and MALDI-TOF/TOF Metabolomic GC-MS Enzyme activity and activity gels	Differences between peel and juice sacs regarding changes in metabolites and antioxidant activity in response to HT; putrescene high in peel, proline in juice sacs; upregulation of stress/defense proteins; HIGHER SUGARS; Less pathogen infection and CI after 30 and 60 days at 5 °C	Perotti *et al.*, 2011[38]
Citrus paradise cv. Star Ruby	HWB 20 s 62 °C	Transcriptomic cDNA libraries Northern blot analysis	Stress-related, antioxidant and lipid modification genes analysed. HT primed genes to be expressed in cold storage. CI reduced after 8 and 14 weeks at 2 °C	Sapitnitskaya*et al.*, 2006[37]
Peach				
*Prunuspersica*L. Batsch cv. Huiyulu	HWT 10 min 48 °C	Proteomic 2-DE and MALDI-TOF/TOF	Induction of stress/defense and cell and protein structure proteins	Zhang *et al.*, 2011[42]

*Prunuspersica*L. Batsch cv. Dixiland	HAT 3 days 39 °C	Metabolic GC-MS qRT-PCR	Sugar and sugar alcohols increased in HT Amino-acid precursors of phenyl-propanoid pathway modified Both common and specific responses to heat and cold	Lauxmann *et al.*, 2014[46]
*Prunuspersica*L. Batsch cv. Dixiland	HAT 3 days 39 °C	Transcriptomic Differential display qRT-PCR	127 differentially expressed unigenes—47% heat-induced and 36% heat-repressed 70% of heat responsive genes found to be responsive to cold in *Arabidopsis*	Lauxmann *et al.*, 2012[45]
*Prunuspersica*L. Batsch cv. Dixiland	HAT 3 days 39 °C	Proteomic of apoplastic proteins 2D-DIGE MS/MS qRT-PCR	Cell wall modifiying genes less expressed in HT fruit Increase of DUF642 and glyceraldehydes-3-phosphate dehydrogenase proteins in HT apoplast	Bustamanteet *al.*, 2012[44]
*Prunuspersica*L. Batsch cv. Dixiland	HAT 3 days 39 °C	Proteomic 2D-DIGE MALDI TOF-TOF Enzyme assays	Increase in stress/defense proteins, decrease in polyphenol oxidase	Lara *et al.*, 2009[43]
Tomato				

*Solanium lycopersicum*L. cv. Micro-Tom	HWT 7 min 40 °C	Metabolic GC-MS	PCA separated treatments. HWT was different from chilled Arabinose, fructose-6-phosphate, valine, shikimic acid associated with HWT induced resistance to CI	Luengwilai *et al.* 2012[50]
*Solanium lycopersicum*L. cv. Messina	HAT 12 h 38 °C	Transciptomic qRT-PCR Metabolic HPLC for amino acids, enzyme assay	Arginase transcript and content correlated with HT inhibition of CI. HT increased antioxidant enzyme activity. Putrescene and proline levels higher after HT	Zhang *et al.*, 2013[51]
Potato				
Solanium tuberosum L.	HAT 7 days 34 °C	Transcriptomic cDNA microarray qRT-PCR	Stress transcription factors and cell proliferation and differentiation induced by HT	Ginzberg *et al.*, 2009[48]

CITRUS

Citrus fruits are susceptible to chilling injury when exposed to temperatures lower than 2–5 °C. High temperature treatments using HAT[28] and HWT[29, 30] can delay the appearance of chilling injury and inhibit fungal infection. Sala and Lafuente[28] found a correlation between catalase (CAT) activity and resistance to chilling injury when mandarins were stored at 2 °C following a 3 min HWT at 53 °C or 3-day HAT at 38 °C hot air. They did not find induction of ascorbate peroxidase (APX), glutathione reductase or superoxide dismutase (SOD) in the heated fruit. This treatment of HAT can also increase polyamine concentration and the increase correlated with lower chilling sensitivity of mandarins.[31]

The most common treatment for citrus is a HWT for 2–3 min at temperatures of 50–53 °C to control fungal development. Yun *et al*[32] conducted a proteomic and metabolomic study of mandarin peel during storage after a 2 min dip at 52 °C. This treatment successfully suppressed *Penicillium italicum* development, and reduced chilling injury during storage. Two-dimensional gel electrophoresis (2-DE) found 50 proteins (out of 600 detected spots) that were altered at least 1.5-fold between heated and control fruit during storage ranging from 1 to 39 days after treatment. These proteins fell into the categories of glycolysis and TCA cycle, redox regulation, stress response and protein folding. The stress proteins were higher in the heated fruit and included HSPs and PR proteins, and some remained higher than control for up to 39 days after treatment. The ROS metabolism-related proteins that were found differentially in the 2-DE gels were lower during storage in heat treated fruit, and included a number of oxidoreductases and Cu/Zn SOD. However, a chloroplast APX was induced by the heat treatment and remained high in storage. CAT was not found differentially in this study. Although the ROS enzymes were not higher in the heated fruit, H_2O_2decreased in the heat treated fruit during storage, while it increased in the control fruit.

The metabolomic study identified 62 metabolites which were grouped into alcohols, amino acids, sugars, organic acids and fatty acids.[32] Most sugars and fatty acids increased in the peel after HWT, but the sugar levels returned to control levels during storage. Organic acids were lower immediately after HWT and then recovered to a higher level than control fruit during storage. Amino acids were also decreased by HWT, but remained lower than control even during storage. An exception was ornithine which was 2.5 times higher than control fruit. This amino acid is required for the synthesis of polyamines and alkaloids, which contribute to abiotic stress tolerance in plants.[33]

Immediately after the HWT, many flavonoids were much higher than in control fruit, including quercitin, hesperetin, naringenin and rutin. Most of these have antioxidant activity and may have contributed to the decrease in H_2O_{2-}.[34,35] Some also have antimicrobial properties.[36] Most increases in these compounds disappeared during the first few days of storage.

A transcriptomics approach[37] found many similarities to the proteomic results of Yun *et al.*[32] The treatment was a HWB of grapefruit for 20 s at 62 °C, and the fruit were stored at 2 °C for 8 and 14 weeks. There was 80% decrease in chilling injury in the fruit after storage. cDNA libraries were constructed and sequenced and then subtraction analysis was performed. Eleven different cDNAs were evaluated after the HWB and cold storage. These were six stress-responsive, three antioxidant and two lipid modification genes chosen for verification with RNA Northern blots. The HSPs were highly expressed following the HWB as were other stress genes such as dehydrin and glucanase. Lipid modifying genes, fatty acid desaturase, lipid transfer protein and lipase were induced by HWB and decreased in storage. The antioxidant genes of SOD and CAT increased during cold storage. Many of the gene expression patterns showed a priming effect from the HWB; the expression was not higher immediately after the treatment, but increased over time during exposure to chilling.

In contrast to the HWT and HWB which are short heat treatments, antioxidants were increased by in oranges by a HAT for 48 h at 37 °C followed by storage at 5 °C for 30 and 60 days.[38] Proteomics was conducted on both flavedo (peel) and juice sacs, and activity gels were also run for SOD, POD and APX. The activity of all three enzymes increased after HAT and remained higher than control fruit during storage. CAT and glutathione reductase activities did not change due to HAT. From the proteomic 2-DE results of juice sacs, 61 differential protein spots (out of 500 spots) were detected, either after the HAT or after 30 or 60 days of storage. The classes of proteins increased by the HAT were defense, including both HSPs and PR proteins, and metabolism. However, after 60 days at 5 °C, the differential spots found between HAT or control fruit were all higher in the control fruit. Metabolic analysis was made of both flavedo and juice sacs after treatment and after 60 days of storage and 28 compounds were analyzed. Sugars were higher in the HAT at both time points. The main organic acids in the juice, citric and malic, were not affected by the HAT, while they decreased in the flavedo initially and then recovered during storage as occurred in the HWT of Yun *et al.*[32]Glycerol and putrescine increased in the HAT flavedo, while proline was very high in the fruit sacs. Proline has been found to be correlated in many plants with adaptation to cold temperatures.[39]Putrescine, one of whose precursors ornithine was high in HWT mandarins,[32] was also found by Gonzalez Aguilar *et al.*[31] to increase in mandarins during HAT. Putrescine is essential to counteract cold stress in *Arabidopsis*40 and to allow ABA increases in cold stressed plants.[41]

PEACH

Peach fruit has been subjected to both HWT and HAT to study molecular changes in the fruit. A HWT of 48 °C for 10 min followed by shelf life was used to examine proteome changes during ripening.[42]Using 2-DE, 600 protein spots were detected and 35 differentially expressed spots were analyzed. Almost half of the proteins identified were stress and defense, including antioxidants. As in the study of HWB of grapefruits,[37] in many cases, the increase in protein level occurred with a delay of a day or two after the treatment. Both APX and DHAR were higher in protein abundance 1 and 3 days after the treatment, as were two HSPs.

A series of papers have been published detailing changes in peach fruit due to a HAT of 39 °C for 3 days.[43, 44, 45, 46] This treatment was found to delay internal breakdown development in stored peaches, although it enhanced the red coloration in both peel and flesh.[47] A study by Lara *et al.*[43]followed the recovery of the fruit from the heat stress, and the resumption of ripening which was inhibited during HAT, particularly softening and ethylene gene expression.

D-glucose, *D*-fructose, sucrose, sorbitol, malic acid and citric acid were determined using enzymatic kits. The analysis of these metabolites showed that, as in citrus, sugars (glucose, fructose, sorbitol) were higher than in control fruit during and following the HAT, while organic acids (citric and malic) were lower. Sucrose degradation, and glycolic- and fermentation-related enzyme activity and transcript levels by qRT-PCR were analyzed after HAT and 3 days later. Enzyme activity and expression levels of invertases (both acid and neutral), pyruvate kinase and pyruvate decarboxylase, ATP-dependent and PPi-dependent phosphofructokinase were higher in HAT fruit 3 days after the treatment than in control fruit, though not at the end of the heat treatment. A 2D-DE examination of peach proteins detected 600 spots, and 57 were significantly affected by temperature. Forty-four differential proteins were identified. Twenty-seven percent of the identified proteins corresponded to HSPs. Some of these appeared during the HAT, while others increased following the treatment. An examination of the extracellular proteins of the fruit at harvest found the most abundant proteins were proteases, peroxidases, a porin and a small

HSP.[44] In HAT fruit, a glyceraldehydes-3-phosphate dehydrogenase was abundant, and was localized to the cell wall by immunolocalization. qRT-PCR expression analysis of cell wall modifying genes (which were not identified in the proteome study, perhaps because the technique extracted only weakly bound cell wall proteins) had lower expression during HAT, corresponding to inhibited softening.

A transcriptomic and metabolomic study was conducted on the HAT peaches, after they were placed at 0 °C for a 5 days, with the purpose of determining which responses and metabolic pathways were activated in both stresses and which were confined to either heat or cold.[45,46] Differential transcriptome analysis of the peaches found 127 unigenes that were differentially expressed among heated or unheated placed at 20 °C, 60 unigenes were induced by the HAT and 45 unigenes were repressed.[45] Twenty percent of the genes were involved in protein modification, transcription and RNA metabolism. This group may be responsible for the signal transduction leading to the previously reported proteomic changes.[42,43] Another functional group of genes over-represented in induced genes were those responsive to heat, cold or freezing stress. A group of genes that were either induced or repressed by HAT were chosen for further examination and their expression determined in peaches placed at 0 °C. It was found that 75% of the HAT induced and 95% of the HAT genes repressed were activated similarly by a cold treatment. When the unigenes from peach were compared to *Arabidopsis* orthologs, 70% of them were responsive to cold. The conclusion of this study was that the heat treatment activates many of the same genes that cold stress does, and that some of them confer to the fruit resistance to chilling injury.

The metabolomic study identified 47 metabolites.[46] A PCA analysis of these compounds explained 65% of the variance by three principal components. PC1 separated HAT fruit (both at time of treatment and after removal to 20 °C or 2 °C) from all other samples. PC2 separated harvest from all other samples, while PC3 grouped the samples according to temperature regime, with 20 °C, HAT and 0 °C all clustering separately. This is quite different from the transcriptome study where there was a convergence of transcripts activated by both heat and cold stress.

The metabolites fell into the same classes as were analyzed in the citrus studies; sugars, organic acids, amino acids, fatty acids, and also putrescine and urea.[32,38] The peach study did not include analysis of flavonoids as was done in citrus. Changes in the metabolome of peach during and after HAT showed similarity to changes in citrus. The sugars and alcohol sugars were increased in HAT and most remained high when fruit were transferred to 20 °C, though fructose and glucose returned to control fruit levels. Galactinol was the most elevated by HAT and remained high if fruit were placed in 0 °C, but not when they were put at 20 °C. Organic acids tended to decrease in HAT fruit compared to their levels at harvest, similar to their behavior in citrus peel.[32,38] However, when the same time point was compared (the 3 days HAT compared to 3 days 20 °C) citrate was twice as high, while 2-oxoglutarate, malate and glycerate were 50% lower in content. The amino acid levels in citrus were decreased by HWT, while in HAT peach fruit, 10 out of 14 amino acids were modified, with most increasing either during or after the HAT. Two fatty acids, tetradecanoic and octadecanoic showed opposite responses to HAT with the former increasing and the latter decreasing, similarly to their changes in citrus peel.[32] Putrescine, which increased in citrus after HAT,[38] was lower in HAT peach fruit.

POTATO

The transcriptomic study of potato tubers was for a different purpose than the studies on heat treatment of other commodities. Under adverse growing conditions, particularly high temperature, the skin separates from the other layers and russeting occurs, so the study was to determine gene expression that might be involved in this disorder, as opposed to other commodities where a high-temperature treatment was to develop resistance to storage disorders. The outer periderm tissue of the tuber is made up of three tissue types; phellem, phellogen and phelloderm. As phellem cells become suberized and die, they form a protective layer termed ‘skin’. The phelloderm cells are the innermost layer of the periderm and contain actively dividing cells during tuber maturation. A transcriptomic study was made of periderm tissue of potato tubers utilizing a cDNA microarray containing 11412 cDNA clones to identify genes involved in the periderm response to heat stress.[48]Skin and phelloderm were collected separately from tubers exposed to soil temperatures of 33 °C for a week. In skin tissue, 86 genes were increased and 189 repressed by the heat, while in periderm, 167 genes were increased and 91 were repressed. The major functional group that was differentially expressed in both skin and phelloderm was stress and defense related genes. The major upregulated phelloderm genes were HSPs, while in the skin, they were genes involved in the development of protective/symbiotic plant–microbe interactions (including PR proteins). The transcription factors which were differentially expressed in HAT tissue were those for stress response (33%) and cell proliferation and differentiation (28%). The data suggest that during high temperature, increased periderm/skin layers are produced, leading to reduced flexibility and cracking as the tuber continues to grow. The periderm is a protective tissue containing a wide array of plant defense components.[49]The enhanced cell proliferation activity could be part of the response to heat stress, by creating an insulating cover and preventing water loss from the tuber flesh.

TOMATO

Two studies have recently been published on the ability of heat treatment to prevent chilling injury in tomato.[50,51] Early work on tomato showed that high temperature conditioning of tomato fruit before storage at chilling temperatures (<10 °C) mitigated the development of chilling injury.[52] This mitigation is correlated with altered protein synthesis including accumulation of HSPs and inhibition of normal ripening processes.[26,53] However, other processes may be involved in the preconditioning mitigation of chilling injury. Low-temperature stress causes changes in the accumulation of distinct metabolites in many plant tissue, and changes in the levels of a subset of metabolites in response to preconditioning may contribute to induced chilling tolerance.[54] Metabolic profiling was made of Micro-tom tomatoes comparing control fruit to HWT fruit at 40 °C for 7 min and then held at 20 °C or 2.5 °C for 14 days.[50] Sixty-five metabolites were identified from polar extracts of the fruit. PCA analysis explained 75% of the variance, but the best separation among the treatments was on the PC2 axis which just explained 6% of the variance. This indicates that the metabolites in the samples were quite similar. Examining the metabolites which contributed to PC2 found that low temperature increased arabinose, citric acid, dehydroascorbic acid, fructose-6-phosphate, glucose-6-phosphate, rhamnose and valine, and decreased glutamic acid and shikimic acid. HWT caused increases in four sugars, three organic acids, one fatty acid, one amino acid, as well as allantoin and putrescine, and the levels remained high during cold storage. The authors suggest that the increases in arabinose and fructose-6-phosphate which occur in chilled fruit, but not in HWT and chilled fruit may be markers for chilling injury.

Zhang *et al.*[51] investigated the role of the products of the arginine pathway in contributing to resistance to chilling injury. Arginine is a precursor of polyamines, proline and nitric oxide as well as proteins. Arginase catalyzes the hydrolysis of arginine to urea and ornithine, which is the precursor of putrescine and proline. All these compounds have been found to be present at elevated levels in various heated fruit, including citrus and peach.[32,38,46] Tomato fruit HAT treated at 38 °C for 12 h and stored at 2 °C did not develop chilling injury. Transcript level and activity of arginase were induced by HAT. Following HAT, the stored fruit had higher levels of arginine, proline and putrescine, as well as increased activities of the antioxidative enzymes SOD, CAT and APX. Pre-treatment with an arginase inhibitor cancelled these changes evoked by the HAT.

CONCLUSIONS

There are a number of common themes that run through most of the studies reported here. One is that changes induced by high temperature persist for an extended time, sometimes days and sometimes weeks, when the commodity is placed in low temperature following the heat stress. In the case of short times of exposure to high temperature, the molecular changes may actually occur after the heat stress has ended. The changes involve accumulation of stress and defense proteins, including HSPs, PRs, dehydrins, and antioxidant enzymes and compounds (polyphenols). There is also elevation of sugars and accumulation of putrescince (or its precursors) and proline, both of which have been shown to be involved in adaptation to low temperatures. The study of Lauxmann *et al.*[45] showed the commonality between the response to heat stress and that of cold.

The times and temperatures given in the studies are very different, ranging from 33 °C for days to 63 °C for seconds. In all cases, however, similar changes at the molecular level were obtained. Some of the treatments were long enough so that the changes in gene expression, protein accumulation, and metabolic shifts occurred during the treatment, and in other cases, the treatment primed the tissue to produce the changes after the heat stress.

What is clear from the studies is that more information is acquired from the transcriptomics data than either proteomics or metabolomics. This is undoubtedly due in the case of proteomics to both the sensitivity of the analysis and the cost of analyzing the proteins on the gel. In none of the studies have the majority of protein spots on a gel been analyzed, unlike the full annotation given in transcriptomics studies. In the metabolomic studies, often the extraction is of only one type, for example, polar solvent,[50] and so other compounds are not analyzed. A thorough metabolomics study is more complicated and would need a number of extractions and assays.[32] In the future, these techniques will undoubtedly become more standardized and less expensive and will lead to a more thorough understanding of the changes occurring during high temperature treatment.

One area which was not examined thoroughly in any of the studies reported here was an analysis of the transcription factors and signaling pathways or networks that are involved in the tissue response to a heat treatment. Possibly this is because the studies involved the responses at the end of the stress and during subsequent storage, while the signaling would have been maximal at an early stage of the treatment. However, I am certain that the HSFs are not the only transcription factors involved in complicated response of the tissue to heat stress, and elucidation of these pathways will help untangle the responses.

Currently the research in this field is at the stage of accumulating knowledge without regard to practical application. Since there are many responses to heat treatment that are the same in the different fruit that have been studied so far, it may be possible to choose a model fruit, such as tomato, to conduct a meta-analysis. Ideally, such a study would include different heat treatments, different stages of fruit ripeness and an analysis of all the changes from gene activation through protein and metabolite production. One potential practical outcome could be the development of markers, either a metabolite, stress compound or transcription level of a gene, which would indicate the efficacy of a treatment. As mentioned in the introduction, treatments are developed empirically by trial and error. A more precise measurement of how the plant tissue is responding to the high temperature could help in developing better treatments.

REFERENCES

1. Lurie S. Postharvest heat treatments. *Postharvest Biol Technol* 1998; 14: 257–269.
2. Lurie S. Postharvest heat treatments of horticultural crops. *Hort Rev* 1998; 22: 91–122.
3. Paull RE, Chen NJ. Heat treatment and fruit ripening. *Postharvest Biol Technol* 2000; 21: 21–37.
4. Fallik E. Prestorage hot water treatments (immersion, rinsing and brushing). *Postharvest Biol Technol* 2004; 32: 125–134.
5. Tang J, Mitcham E, Wang S, Lurie S. *Heat Treatments for Postharvest Pest Control: Theory and Practice*. Oxon: CABI International, 2007.
6. Ghasemnezhad M, Marsh K, Shilton R, Babbalar M, Woolf A. Effect of hot water treatments on chilling injury and heat damage in 'satsuma' mandarins: antioxidant enzymes and vacuolar ATPase and pyrophosphatase. *Postharvest Biol Technol* 2008; 48: 364–371.
7. Kotak S, Larkindale J, Lee U, von Koskull-Doring P, Vierling E, Scharf KD. Complexity of the heat stress response in plants. *Curr Opin Plant Biol* 2007; 10: 310–316.
8. Mittler R, Finka A, Goloubinoff P. How do plants feel the heat? *Trends Biochem Sci* 2012; 37: 118–125.
9. McClung CR, Davis SJ. Ambient thermometers in plants: from physiological outputs towards mechanisms of thermal sensing. *Curr Biol* 2010; 20: 1086–1092.
10. Ruelland E, Zachowski A. How plants sense temperature. *Environ Exp Bot* 2010; 69: 225–232.
11. Suzuki N, Koussevitzki S, Mittler R. ROS and redox signaling in the response of plants to abiotic stress. *Plant Cell Environ* 2011;

35: 259–270.

12. Konigshofer H, Thromballa HW, Loppert HG. Early events in signaling high-temperature stress in tobacco BY2 cells involve alterations in membrane fluidity and enhanced hydrogen peroxide production. *Plant Cell Environ* 2008; 31: 1771–1780.
13. von Koskull-Doring P, Scharf KD, Nover L. The diversity of plant heat stress transcription factors. *Trends Plant Sci* 2007; 12: 452–457.
14. Saidi Y, Finka A, Muriset M *et al.* The heat shock response in moss plants is regulated by specific calcium permeable channels in the plasma membrane. *Plant Cell* 2009; 21: 2829–2843.
15. Mishkind M, Vermeer J, Darwish E, Munnik T. Heat stress activates phospholipase D and triggers PIP2 accumulation at the plasma membrane and nucleus. *Plant J* 2009; 60: 10–21.
16. Zheng SZ, Liu YL, Li B, Shang ZL, Zhou RG, Sun DY. Phosphoinositide-specific phospholipase C9 is involved in the thermotolerance in *Arabidopsis. Plant J* 2011; 69: 689–700.
17. Wang L, Yu X, Wang H *et al.* A novel class of heat-responsive small RNAs derived from the chloroplast genome of Chinese cabbage (*Brassica rapa*). *BMC Genomics* 2011; 12: 289.
18. Yu X, Wang H, Lu Y *et al.* Identification of conserved and novel microRNAs that are responsive to heat stress in *Brassica rapa. J Exp Bot* 2012; 63: 1025–1038.
19. Neta-Sharir I, Isaacson T, Lurie S, Weiss D. Dual role for tomato heat shock protein 21: protecting photosystem II from oxidative stress and promoting color changes during fruit maturation. *Plant Cell* 2005; 17: 1829–1838.
20. Larkindale J, Vierling E. Core genome responses involved in acclimation to high temperature.*Plant Physiol* 2008; 145: 748–761.
21. Qin D, Wu H, Peng H *et al.* Heat stress-responsive transcriptome analysis in heat susceptible and tolerant wheat (*Triticum aestivum* L.) by using Wheat Genome Array. *BMC Genomics*2008; 9: 432.
22. Zhang JH, Huang WD, Pan QH, Liu Y. Improvement of chilling tolerance and accumulation heat shock proteins in grape berries (*Vitis vinifera* cv. Jingxiu) by heat pretreatment.*Postharvest Biol Technol* 2005; 38: 80–90.

23. Yi SY, Sun AQ, Sun Y, Yang JY, Zhao CM, Liu J. Differential regulation of Lehsp23.8 in tomato plants: analysis of a multiple stress inducible promoter. *Plant Sci* 2006; 171: 398–407.

24. Sevillano L, Mar Sola M, Vargas AM. Induction of small heat-shock proteins in mesocarp of cherimoya fruit (*Annona cherimola* Mill.) produces chilling tolerance. *J Food Biochem* 2010;34: 625–638.

25. He LH, Chen JY, Kuang JF *et al.* Expression of three sHSP genes involved in heat pretreatment-inducing chilling tolerance in banana fruit. *J Sci Food Agric* 2012; 92: 1924–1930.

26. Sabehat A, Weiss D, Lurie S. The correlation between heat-shock protein accumulation and persistence and chilling tolerance in tomato fruit. *Plant Physiol* 1996; 110: 531–537.

27. Ferguson IB, Lurie S, Bowen JH. Protein synthesis and breakdown during heat shock of cultured pear (*Pyrus communis*) cells. *Plant Physiol* 1994; 104: 1429–1437.

28. Sala JM, Lafuente MT. Catalase enzyme is related to tolerance of mandarin fruit to chilling.*Postharvest Biol Technol* 2000; 20: 81–89.

29. Schirra M, Agabbio M, D'Hallewin G, Pala M, Ruggiu R. Response of Tarocco orange of picking date, postharvest hot water dips, and chilling storage temperature. *J Agric Food Chem*1997; 45: 3216–3220.

30. Schirra M, D'Hallewin G. Storage performance of Fortune mandarins following hot water dips.*Postharvest Biol Technol* 1997; 10: 229–237.

31. Gonzalez-Aguilar GA, Zacarias L, Perez-Amador MA, Carbonell J, Lafuente MT. Polyamine content and chilling susceptibility are affected by seasonal changes in temperature and by conditioning temperature in cold-stored 'Fortune' mandarin fruit. *Physiol Plant* 2000; 108: 140–146.

32. Yun Z, Gao H, Liu P *et al.* Comparative proteomic and metabolomic profiling of citrus fruit with enhancement of disease resistance by postharvest heat treatment. *BMC Plant Biol* 2013; 13: 44.

33. Kalamaki MS, Alexandrou D, Lazari D *et al.* Over-expression of a tomato *N*-acetyl-L-glutamate synthase gene (SlNAGS1) in *Arabi dopsis thaliana* results in high ornithine levels and increased tolerance in salt and drought stresses. *J Exp Bot* 2009; 60: 1859–

1871.

34. Nijveldt RJ, van Nood E, van Hoorn DE, Boelens PG, van Norren K, van Leeuwen PAM.Flavonoids: a review of probably mechanisms of action and potential applications. *Am J Clin Nutr* 2001; 74: 418–425.
35. Saviranta NM, Veeroos L, Granlund JL, Hassinen VH, Kaarniranta K, Karjalainen RI. Plant flanonol quercetin and isoflavone biochanin A differentially induce protection agains oxidative stress and inflammation in ARPE-9 cells. *Food Res Int* 2011; 44: 109–113.
36. Treutter D. Significance of flavonoids in plant resistance: a review. *Environ Chem Lett* 2006;4: 147–157.
37. Sapitnitskaya M, Maul P, McCollum GT *et al.* Postharvest heat and conditioning treatment activate different molecular responses and reduce chilling injuries in grapefruit. *J Exp Bot*2006; 57: 2943–2953.
38. Perotti VE, del Vecchio HA, Sansevich A *et al.* Proteomic, metabolomic, and biochemical analysis of heat treated Valencia oranges during storage. *Postharvest Biol Technol* 2011; 62: 97–114.
39. Chen WP, Li PH. Membrane stabilization by abscisic acid under cold aids proline in alleviating chilling injury in maize (*Zea mays* L.) cultured cells. *Plant Cell Environ* 2002; 25: 955–962.
40. Cuevas JC, Lopez-Cobollo R, Alcazar R *et al.* Putrescine is involved in Arabidopsis freezing tolerance and cold acclimation by regulating abscisic acid levels in response to low temperature. *Plant Physiol* 2008; 148: 1094–1105.
41. Cuevas JC, Lopez-Cobollo R, Alcazar R *et al.* Putrescine as a signal to modulate the indispensable ABA increase under cold stress. *Plant Sign Behav* 2009; 4: 219–220.
42. Zhang L, Yu Z, Jiang L, Jiang J, Luo H, Fu L. Effect of postharvest heat treatment on proteome change of peach fruit during ripening. *J Proteomics* 2011; 74: 1135–1149.
43. Lara MV, Borsani J, Budde CO *et al.* Biochemical and proteomic analysis of 'Dixiland' peach fruit (*Prunus persica*) upon heat treatment. *J Exp Bot* 2009; 60: 4315–4333.
44. Bustamante CA, Budde CO, Borsani J *et al.* Heat treatment of peach fruit: modifications in the extracellular compartment

and identification of novel extracellular proteins. *Plant Physiol Biochem* 2012; 60: 35–45.

45. Lauxmann MA, Brun B, Borsani J *et al.* Transcriptomic profiling during the post-harvest of heat-treated Dixiland *Prunus persica* fruits: common and distinct response to heat and cold.*PLoS ONE* 2012; 7: e51052.
46. Lauxmann MA, Borsani J, Osorio S *et al.* Deciphering the metabolic pathways influencing heat and cold responses during post-harvest physiology of peach fruit. *Plant Cell Environ* 2014;37: 601–616.
47. Murray R, Lucarngeli C, Polenta G, Budde C. Combined pr-storage heat treatment and controlled atmosphere reduced internal breakdown of 'Flavortop' peach. *Postharvest Biol Technol* 2007; 44: 116–121.
48. Ginzberg I, Barel G, Ophir R *et al.* Transcriptomic profiling of heat-stress response in potato periderm. *J Exp Bot* 2009; 60: 4411–4421.
49. Barel G, Ginzberg I. Potato skin proteome is enriched with plant defense components. *J Exp Bot* 2008; 59: 3347–3357.
50. tomato (*Solanium lycopersicum* L.) fruit is altered by chilling and protective heat-shock treatments as shown by GC-MS metabolic profiling. *Postharvest Biol Technol* 2012; 63: 116–122.
51. Zhang X, Shen L, Li F, Meng D, Sheng J. Arginase induction by heat treatment contributes to amelioration of chilling injury and activation of antioxidant enzymes in tomato fruit. *Postharvest Biol Technol* 2013; 79: 1–8.
52. Lurie A, Klein JD. Acquisition of low-temperature tolerance in tomatoes by exposure to high temperature stress. *J Am Soc Hort Sci* 1991; 116: 1007–1012.
53. Saltveit MES. Influence of heat shocks on the kinetics of chilling-induced ion leakage from tomato pericarp discs. *Postharvest Biol Technol* 2005; 36: 87–92.
54. Sevillano L, Sanchez-Ballesta MT, Romojaro F, Flores FB. Physiological, hormonal and molecular mechanisms regulating chilling injury in horticultural species. Postharvest technologies applied to reduce its impact. *J Sci Food Agric* 2009; 89: 555–573.

Chapter 3

An Economic and Ecological Perspective of Ethanol Production from Renewable Agro Waste: A Review

Latika Bhatia, Sonia Johri, and Rumana Ahmad

Department of Life Sciences, Institute of Technology & Management University, Gwalior, Madhya Pradesh, 475001, India

ABSTRACT

Agro-industrial wastes are generated during the industrial processing of agricultural products. These wastes are generated in large amounts throughout the year, and are the most abundant renewable resources on earth. Due to the large availability and composition rich in compounds that could be used in other processes, there is a great interest on the reuse of these wastes, both from economical and environmental view points. The economic aspect is based on the fact that such wastes may be used as low-cost raw materials for the production of other value-added compounds, with the expectancy of reducing the production costs. The environmental concern is because most of the agro-industrial wastes contain phenolic compounds and/or other compounds of toxic potential; which may cause deterioration of the environment when the waste is discharged to the nature. Although the production of bioethanol offers many benefits, more research is needed in the aspects like feedstock preparation, fermentation technology modification, etc., to make bioethanol more economically viable.

INTRODUCTION

Since the late 19th century, the mean temperature on earth has increased with 0.8°C and the major part of this increase is likely due to anthropogenic emissions of greenhouse gases. Carbon dioxide is the type of greenhouse gas with largest emission and this originates from the combustion of fossil fuels as coal, oil and natural gas (Sun and Cheng 2002). In USA, transportation accounts for 30% of the total energy consumption. Burning fossil fuels such as coal and oil releases CO_2, which is a major cause of global warming. With only 4.5% of the world's population, the US is responsible for about 25% of global energy consumption and 25% of global CO_2 emissions. The average price of gasoline in 2005 was $2.56 per gallon, which was $0.67 higher than the average price of gasoline in the previous year. Yet in June 2008, the average price of gasoline in the US reached $4.10 per gallon (Kumar et al. 2009).

Soaring oil prices associated with concerns of climate change and national energy security are driving us to utilize sustainable alternative energy sources, such as solar energy, nuclear energy, wind energy, hydropower, tidal energy, and so on (Lynd et al. 2003). With the inevitable depletion of world's energy supply, there has been an increasing interest worldwide in alternative sources of energy. Unlike fossil fuels, ethanol is a renewable energy source produced through fermentation of sugars and used as a partial gasoline replacement in a few countries of the world (Sharma et al.2007).

Bioethanol market is expected to reach 100 x10^9 liters in 2015. The largest producers in the world are the US, Brazil, and China. In 2009, US produced 39.5x10^9 liters of ethanol using corn as a feedstock while the second largest producer, Brazil, created about 30x10^9 liters of ethanol using sugarcane. China is a country that has invested much in the production of ethanol, and is nowadays one of the largest ethanol producers (Ivanova et al. 2011).

Ethanol contains 35% oxygen, which results in a complete combustion of fuel and thus lowers the emission of harmful gases. Moreover, ethanol production uses energy from renewable sources only; hence, no net carbon dioxide is added to the environment, thus reducing green-house gas emissions. It has also been well established now that ethanol increases the octane number, decreases the Reid vapor pressure and produces fuel with clean burning characteristics (Dhillon et al. 2007). Moreover, neat (unblended) ethanol can be burned with greater efficiency, and is thought to produce smaller amounts of ozone precursors (thus decreasing urban air pollution), and is particularly beneficial with respect to low net CO_2 put into the atmosphere.

The increasing demand for various industrial purposes such as alternative source of energy, industrial solvents, cleansing agents and preservatives, has necessitated increased production of ethanol (Brooks 2008). Furthermore, ethanol by fermentation offers a more favorable trade balance, enhanced energy security, and a major new crop for a depressed agricultural economy. Ethanol is considerably less toxic to humans than is gasoline (or methanol). Ethanol also reduces smog formation because of low volatility; its photochemical reactivity and low production of combustion products. Furthermore, low levels of smog-producing compounds are formed by its combustion (Wyman and Hinman 1990). In addition, the low flame temperature of ethanol results in good engine performance.

Currently, bioethanol is being commercially produced only from edible feedstock such as corn-starch and sugarcane juice. The European Union (EU) had established a goal of 5.75% biomass-derived transportation fuels by December, 2010. The use of fuel ethanol has been quite successful in Brazil, where it is being produced at a very low cost by fermentation of sugarcane. In the US, corn is the dominant biomass feedstock for production of ethanol, and in the EU, straw and other agricultural wastes are the preferred types of biomass for ethanol production (Raposo et al. 2009). These bioethanol production systems pose a concern about competition with food and feed supplies. To avoid this competition, bioethanol production from non edible lignocellulosic biomass such as wheat straw, rice straw, bagasse, corn stover, wood, peels of fruits and vegetables is attracting keen interest.

The current production and use of bioethanol processes are a starting point. It is our belief that the next generational change in the use of bioresources will come from a total integration of innovative plant resources, synthesis of biomaterials, and generation of biofuels and biopower (Figure 1).

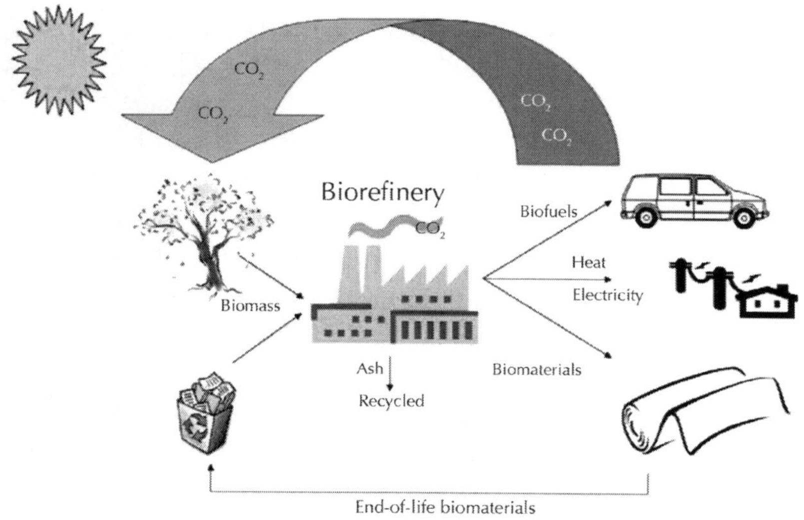

Figure 1: The fully integrated agro-biofuel-biomaterial-biopower cycle for sustainable technologies. (Ragauskas et. al Ragauskas et al. 2006).

The present review is a concise overview of current and latest developments in ethanol production with special emphasis on the choice of lignocellulosic substrates, pretreatment methods and types of microorganisms that have been used for optimal, ecological and economic production of ethanol. Also reviewed are the different fungal and bacterial lignocellulolytic enzymatic systems including the current status of the technology for bioconversion of lignocellulose residues by microorganisms (particularly yeasts and fungi), with focus on the most economical and eco-friendly method for ethanol production.

Lignocellulosic Biomass

Lignocellulose is a renewable organic material and is the major structural component of all plants. Lignocellulose consists of three major components:

- Cellulose, the major constituent of all plant material and the most abundant organic molecule on earth, is a linear biopolymer of anhydroglucopyranose-molecules, connected by β-1, 4-glycosidic bonds. Cellulose or β-1-4-glucan is a linear polysaccharide polymer of glucose made of cellobiose units. The cellulose chains are packed by hydrogen bonds in so-called 'elementary microfibrils'. These fibrils are attached to each other by hemicelluloses, amorphous polymers of different sugars as well as other polymers such as pectin, and covered by lignin. The microfibrils are often associated in the form of bundles or macrofibrils. This special and complicated structure makes cellulose resistant to both biological and chemical treatments. (Delmer and Amor 1995, Morohoshi 1991, Ha et al. 1998).
- Hemicellulose, the second most abundant component of lignocellulosic biomass, is a heterogeneous polymer of pentoses (including xylose and arabinose), hexoses (mainly mannose, less glucose and galactose) and sugar acids. Hemicellulose is less complex, its concentration in lignocellulosic biomass is 25 to 35% and it is easily hydrolysable to fermentable sugars (Saha et al. 2007). The dominant sugars in hemicelluloses are mannose in softwoods and xylose in hardwoods and agriculture residues (Persson et al. 2006, Lavarack et al. 2002, Balan et al.2009).
- Lignin, the third main heterogeneous polymer in lignocellulosic residues, generally contains three aromatic alcohols including coniferyl alcohol, sinapyl and *p*-coumaryl. Lignin serves as a sort of 'glue' giving the biomass fibers its structural strength. Lignin acts as a barrier for any solutions or enzymes by linking to both hemicelluloses and cellulose and prevents penetration of lignocellulolytic enzymes to the interior lignocellulosic structure. Not surprisingly, lignin is the most recalcitrant component of lignocellulosic material to degrade (Zaldivar et al. 2001, Hamelinck et al. 2005).

Lignocellulose Substrates Used for Ethanol Production

Sweet sorghum bagasse can be converted efficiently into fermentable sugars (and is a new potential raw material for fuel ethanol production) by SO_2 catalyzed steam pretreatment at 190°C for 10 min or 200°C for 5 min followed by enzymatic hydrolysis with a result of 89-92% glucan conversion (Sipos et al. 2009). Hemp and ensiled hemp can be converted into ethanol with steam pretreatment (2% SO_2 catalyst, 210°C for 5 min) followed by simultaneous saccharification and fermentation at high solid loading (7.5% water insoluble solids[WIS]) with a result of 171–163 g ethanol/kg raw material (Sipos et al. 2010). In Brazil, ethanol is usually produced from cane juice, whereas in USA, starch-crops such as corn are usually used for ethanol production (Sanchez2009). Using sugars or corn as the main source for ethanol production caused a great deal of controversy due to its effect on food production and costs, which has made it difficult for ethanol to become cost competitive with fossil fuels. These concerns became a driving force in the generation of new biofuel research using lignocellulosic wastes produced by many different industries. Apart from corn and cane juice; wheat, oat and barley straw has also been routinely used to produce up to 0.52 million gallons of ethanol per year (Hahn et al. 2006). China is the world's largest sweet potato (*Ipomoea batatas* Lam.) producer (accounting for 85% of global production), with the output exceeded 100 M tons in 2005 (Lu et al. 2006). Zhang et al. (2011) reported sweet potato as an attractive feedstock for bioethanol production from both economic and environment friendly standpoints.

Lignocellulosic wastes are produced in large amounts by different industries including forestry, pulp and paper, agriculture and food, in addition to different wastes from municipal solid waste (MSW), and animal wastes (Sims 2003, Kim and Dale 2004, Kalogo et al. 2007, Champagne 2007, Wen et al. 2004). Those derived from agricultural activities include materials such as straw, stem, stalk, leaves, husk, shell, peel, lint, seed/stones, pulp or stubble from fruits, legumes or cereals (rice, wheat, corn, sorghum, barley), bagasses generated from sugarcane or sweet sorghum milling, spent coffee grounds, brewer's spent grains, and many others. These potentially valuable materials were treated as waste in many countries in the past, and still are today in some developing countries, which raises many environmental concerns (Palacios-Orueta et al. 2005). Significant efforts, many of which have been successful, have been made to convert these lignocellulosic residues to valuable products such as biofuels, chemicals and animal feed (Howard et al. 2003). Banana peel, an agro waste can be used as a substrate for ethanol production owing to its rich carbohydrate, crude proteins and reducing sugars. Moreover, banana peels are affordable and renewable low cost raw material which makes it potential feedstock for ethanol production (Bhatia and Paliwal 2010). Similarly pineapple is the second harvest of importance after bananas, contributing to over 20% of the world production of tropical fruits (Coveca 2002). Thailand, Philippines, Brazil and China are the main pineapple producers in the world supplying nearly 50% of the total output. Other important producers include India, Nigeria, Kenya, Indonesia, México and Costa Rica and these countries provide most of the remaining fruit available (50%). Isitua and Ibeh 2010 assayed the feasibility of obtaining ethanol from pineapple waste with the purpose of obtaining a valuable product from the residues of the juice and canning industries.

Large volume of bagasse is generated during sugarcane processing. Agricultural profitability and environmental protection issues are associated with disposal of bagasse. In recent years, potential efforts have been directed towards the utilization of cheap renewable agricultural resources, such as sugarcane bagasse as alternative substrate for ethanol production (Bhatia and Paliwal 2011). Rice is the major crop grown worldwide with an annual productivity around 800 million metric tonnes that corresponds with the large production of rice straw. In search for viable alternatives of biofuels, paddy straw has been pursued as suitable lignocellulosic waste for ethanol production (Wati et al. 2007).

Feasibility of lignocellulosic material for ethanol production has been explored around the world depending upon availability. Production of ethanol from wheat straw, one of the most abundant agricultural wastes, has been extensively studied (Ballesteros et al. 2004, Curreli et al. 2002, Curreli et al. 1997, Talebnia et al. 2010). The average yield of wheat straw is 1.3–1.4 lb per lb of wheat grain (Montane et al. 1998). According to Ballesteros et al. 2006, under the 60% ground cover practice, about 354 millions of tons of wheat straw could be available globally and could produce 104 GL of bioethanol. Europe production would account for about 38% of this world bioethanol capacity. In Spain, grain industry generates important amounts of wheat straw, a part of which is used as bedding straw and the remainder is burned or left on the land to fertilize the soil. Bioconversion of this residue to fuel ethanol would provide an attractive possibility to boost the development of biofuels in a sustainable way.

Overview of Lignocellulosic Fermentation

Schematic picture for the conversion of lignocellulosic biomass to ethanol, including the major steps can been seen in Figure 2. Pretreatment of the lignocellulosic residues is necessary because hydrolysis of non-pretreated materials is slow, and results in low product yield. Some pretreatment methods increase the pore size and reduce the crystallinity of cellulose (Dawson and Boopathy2007). Pretreatment also makes cellulose more accessible to the cellulolytic enzymes, which in return reduces enzyme requirements and, thus, the cost of ethanol production. The pretreatment not only enhance the bio-digestibility of the wastes for ethanol production, but also results in enrichment of the difficult biodegradable materials, and improves the yield of ethanol from the wastes.

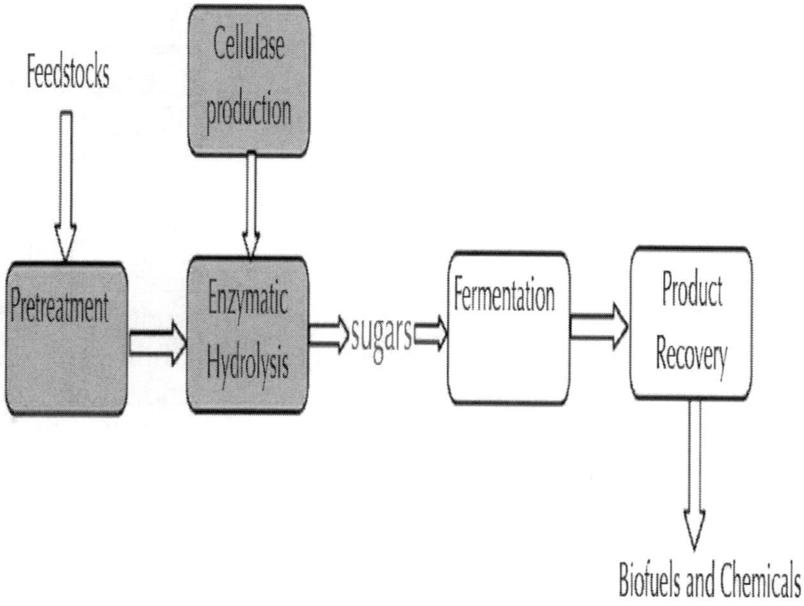

Figure 2: Major steps involved in the conversion of lignocellulosic biomass to ethanol (Dashtban et al.2009).

Post pretreatment, the recalcitrant lignocellulosic biomass becomes susceptible to acid and/or enzymatic hydrolysis as the cellulosic microfibrils are exposed and/or accessible to hydrolyzing agents (Jacobsen and Wyman 2000). In the pretreatment process, small amounts of cellulose and most of hemicellulose is hydrolyzed to sugar monomers; mainly D-xylose and D-arabinose. The pretreated biomass is then subjected to filtration to separate liquids (hemicellulose hydrolysate) and solid (lignin and cellulose). After detoxification, the liquid is sent to a xylose (pentose) fermentation column for ethanol production. Solids are subjected to hydrolysis (also called second stage hydrolysis). This process is mainly accomplished by enzymatic methods using cellulases. Mild acid hydrolysis using sulfuric and hydrochloric acids is an alternative procedure (Zhang and Lynd 2004). The hydrolyzed sugars such as D-glucose, D-galactose, and D-mannose, can be readily fermented to ethanol using various strains of *Saccharomyces cerevisae*. The pentoses (D-xylose and D-arabinose) from hemicellulose hydrolysis are not easily utilized by *Saccharomyces*strains; therefore, genetically modified strains of *Pichia stipitis*, *Zymomonas mobilis*, are used for their fermentation. *Candida shehatae* is capable of co-fermenting both pentoses and hexoses to ethanol and other value-added products at high yields (Betancur 2005, Senthilkumar and Gunasekaran 2005).

Numerous pretreatment strategies have been developed to enhance the reactivity of cellulose and to increase the yield of fermentable sugars. Typical goals of pretreatment include:

- Production of highly digestible solids that enhances sugar yields during enzyme hydrolysis, avoidance of degradation of sugars (mainly pentoses) including those derived from hemicelluloses.
- Minimization of formation of inhibitors for subsequent fermentation steps.
- Recovery of lignin for conversion into valuable co-products.
- Cost effectiveness by operating in reactors of moderate size and by minimizing heat and power requirements (Mosier et al. 2005, Sun and Cheng 2007, Yang and Wyman 2008). Figure 3depicts schematic of goals of pretreatment on lignocellulosic material.

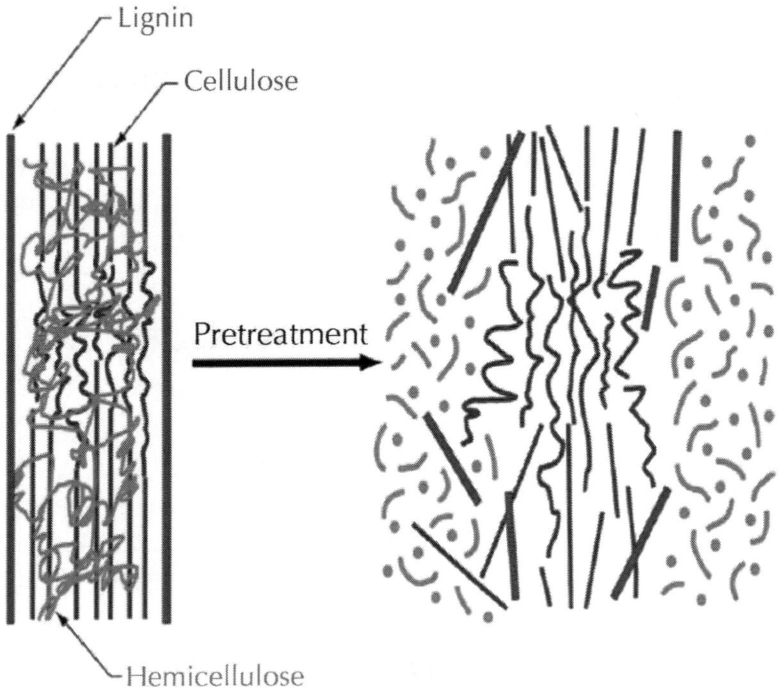

Figure 3: Schematic of goals of pretreatment on lignocellulosic material.

Physical Pretreatments Methods

Physical pretreatments methods such as ball milling and grinding have been used for degradation of lignocelluloses with limited success. This method of pretreatment being cost effective and ecofriendly, and one on which relatively little work has been done and reported, so far, would form one of the thrust areas of future research.

Waste materials can be comminuted by a combination of chipping, grinding and milling to reduce cellulose crystallinity. The size of the materials is usually 10–30 mm after chipping and 0.2–2 mm after milling or grinding. Vibratory ball milling has been found to be more effective in breaking down the cellulose crystallinity of spruce and aspen chips and improving the digestibility of the biomass than ordinary ball milling (Millet et al. 1976). The power requirement of mechanical comminution of agricultural materials depends on the final particle size and the waste biomass characteristics (Cadoche and Lopez 1989).

Pyrolysis has also been used for pretreatment of lignocellulosic materials. When the materials are treated at temperatures greater than 300°C, cellulose rapidly decomposes to produce gaseous products and residual char (Kilzer and Broido 1965, Shafizadeh and Bradbury 1979). The decomposition is much slower and less volatile products are formed at lower temperatures.

The efficiency of ultrasound in the processing of vegetal materials has been already proved (Vinatoru et al. 1999). The known ultrasounds benefits, such as swelling of vegetal cells and fragmentation due to the cavitational effect associated with the ultrasonic treatment, act by increasing the yield and by shortening of the extraction time. The effect of ultrasound on lignocellulosic biomass has been employed in order to improve the extractability of hemicelluloses (Ebringerova et al. 2002), cellulose (Pappas et al. 2002), lignin (Sun and Tomkinson 2002) or to get clean cellulosic fiber from used paper (Scott and Gerber 1995) but only few attempts to improve the susceptibility of lignocellulosic materials to biodegradation by using ultrasound have been described. It was found out that ultrasound has a beneficial effect on saccharification processes (Rolz 1986). Sonication has been reported to decrease cellulase requirements by 1/3 to 1/2 and to increase ethanol production from mixed waste office paper by approximately 20% (Wood et al. 1997). It was notice that the effect of ultrasound fragmentation of Avicel (microcrystalline cellulose formed by acid treatment) is similar to that of the enzymes for short incubation intervals (Gama et al. 1997). The time needed for ultrasonic treatment could be reduced when increasing the irradiation power (Imai et al. 2004).

Chemical Pretreatment Methods

Alkaline Pretreatment

Alkaline pretreatment involves the use of bases, such as sodium, potassium, calcium, and ammonium hydroxide, for the pretreatment of lignocellulosic biomass. The use of an alkali causes the degradation of ester and glycosidic side chains resulting in structural alteration of lignin, cellulose swelling, partial decrystallization of cellulose (Cheng et al. 2010, Ibrahim et al. 2011, McIntosh et al. 2010) and partial solvation of hemicelluloses (McIntosh et al. 2010, Sills et al.2011). Sodium hydroxide has been extensively studied for many years, and it has been shown to disrupt the lignin structure of the biomass, increasing the accessibility of enzymes to cellulose and hemicellulose (MacDonald et al. 1983, Soto et al. 1994, Zhao et al., 2008). Another alkali that has been used for the pretreatment of biomass is lime. Lignocellulosic feedstocks that have been shown to benefit from this method of pretreatment are corn stover, switchgrass, bagasse, wheat, and rice straw.

The conditions for alkaline pretreatment are usually less severe than other pretreatments. It can be performed at ambient conditions, but longer pretreatment times are required than at higher temperatures. The advantage of lime pretreatment is that the cost of lime required for pretreatment of a given quantity of biomass is lowest among alkaline treatments. Most commonly used alkali in the alkali pretreatment processes are NaOH and $Ca(OH)_2$. This process results in (i) the removal of all lignin and part of hemicellulose, and (ii) increased reactivity of cellulose in further hydrolysis steps (Hamelinck et al. 2005), especially, enzymatic hydrolysis. Effective removal of lignin minimizes adsorption of enzyme onto lignin and thus allows for effective interactions with cellulose (Aswathy et al. 2010). Between NaOH and $Ca(OH)_2$, pretreatment with $Ca(OH)_2$ is preferable because it is less expensive, more safer as compared to NaOH and it can be easily recovered from the hydrolysate by reaction with CO_2 (Mosier et al. 2005).

Acid Pretreatment Methods

Acid pretreatment involves the use of concentrated and diluted acids to break the rigid structure of the lignocellulosic material. The most commonly used acid is dilute sulphuric acid (H_2SO_4), which has been commercially used to pre-treat a wide variety of biomass types-switchgrass, corn stover, spruce (softwood), and poplar. Acid pretreatment (removal of hemicellulose) followed by alkali pretreatment (removal of lignin) has shown to yield relatively pure cellulose (Wingren et al. 2003, Taherzadeh and Karimi 2008). Strong acid allows complete breakdown of the components in the biomass to sugars, but also requires large volumes of concentrated sulfuric acid and can result in the production of furfural, an inhibitory byproduct (Goldstein and Easter 1992). Dilute acid allows reduced acid concentrations, but requires higher temperatures, and again gives furfural.

A key advantage of acid pretreatment is that a subsequent enzymatic hydrolysis step is sometimes not required, as the acid itself hydrolyses the biomass to yield fermentable sugars (Zhu et al.2009). A mixture of H_2SO_4 and acetic acid resulted in 90% saccharification (DeMoraes-Racha et al. 2010). Hemicellulose and lignin are solubilized with minimal degradation, and the hemicellulose is converted to sugars with acid pretreatment. The major drawback to these acid processes is the cost of acid and the requirement to neutralize the acid after treatment.

Wet Oxidation

Wet oxidation utilizes oxygen as an oxidizer for compounds dissolved in water. Typically, the procedure for wet oxidation consists of drying and milling lignocellulosic biomass to obtain particles that are 2 mm in length, to which water is added at a ratio of 1 L to 6 g biomass. Wet oxidation has been used to fractionate lignocellulosic material by solubilizing hemicellulose and removing lignin (Martin et al. 2007, Banerjee et al. 2009). It has been shown to be effective in pretreating a variety of biomass such as wheat straw, corn stover, sugarcane bagasse, cassava, peanuts, rye, canola, faba beans, and reed to obtain glucose and xylose after enzymatic hydrolysis (Martin et al.2008, Banerjee et al. 2009, Ruffell et al. 2010, Szijarto et al. 2009, Martin and Thomsen 2007). During wet oxidation, lignin is decomposed to carbon dioxide, water and carboxylic acids. Biomass such as straw, reed and other cereal crop residues have a dense wax coating containing silica and protein which is removed by wet oxidation (Schmidt et al. 2002).

Wet oxidation has been combined with other pretreatment methods to further increase the yield of sugars after enzymatic hydrolysis. Combining wet oxidation with alkaline pretreatment has been shown to reduce the formation of byproducts, thereby decreasing inhibition. Bjerre et al. (1996) used wet oxidation and alkaline hydrolysis of wheat straw (20 g straw/l, 170°C, 5–10 min), and achieved 85% conversion yield of cellulose to glucose. Wet oxidation combined with base addition readily oxidizes lignin from wheat straw, thus making the polysaccharides more susceptible to enzymatic hydrolysis. Furfural and hydroxymethylfurfural, known inhibitors of microbial growth when other pretreatment systems are applied, were not observed following the wet oxidation treatment (Azzam 1989).

Green Solvents

Processing of lignocellulosic biomass with ionic liquids (IL) and other solvents has gained importance in the last decade due to the tunability of the solvent chemistry and hence the ability to dissolve a wide variety of biomass types. Ionic liquid (IL) was found to possess a great potential in dissolving cellulose (Swatloski et al. 2002). Ionic liquids are salts, typically composed of a small anion and a large organic cation, which exist as liquids at room temperature and have very low vapor pressure. The chemistry of the anion and cation has been tuned to generate a wide variety of liquids which can dissolve a number of biomass types-corn stover (Cao et al. 2010), cotton (Zhao et al. 2009), bagasse (Wang et al. 2009), switchgrass, wheat straw (Li et al. 2009).

Dadi and coworkers (2007) have studied the enzymatic hydrolysis of Avicel regenerated from two different ILs, 1-*n*-butyl-3-methylimidazolium chloride and 1-allyl-3-methylimidazoliumchloride. Hydrolysis kinetics of the IL-treated cellulose was significantly enhanced. A limitation in using ionic liquids is the fact they tend to inactivate cellulose.

A solvent which has been effective in dissolution of cellulose and has a low vapor pressure similar to that of the ionic liquids is N-methyl morpholine N-oxide (NMMO). NMMO retains all the advantages of the ionic liquids ability to dissolve a variety of lignocellulosic substrates (Kuo and Lee 2009, Shafiei et al. 2010) without the need to chemically modify them and >99% of the solvent can be recovered due to its low vapor pressure (Perepelkin 2007). It is also nontoxic and biodegradable as proven by the work of Lenzing and other researchers (Rosenau et al. 2001). Further research is needed to evaluate and improve the economics of usage of ILs and NMMO for pretreatment of biomass. Pretreatment of lignocellulosic materials with acidified organic solvents (mixture of 80% ethylene glycol, 19.5% water and 0.5% HCl at 178°C for 90 min) has also been successfully used (Yamashita et al. 2010). The advantages of these methods include recovery and recycling of organic solvents as they can be easily distilled out. The disadvantages are that the process requires expensive high pressure equipment. Their performances could be improved by heating, microwave, or sonication (ElSeoud et al. 2007, Zhu et al. 2006a).

Physicochemical Pretreatment Methods

Steam-Explosion

Steam-Explosion pretreatment is one of the most commonly used pretreatment options, as it uses both chemical and physical techniques in order to break the structure of the lignocellulosic material (McMillan 1994). This hydrothermal pretreatment method subjects the material to high pressures and temperatures for a short duration of time after which it rapidly depressurizes the system, disrupting the structure of cellulose microfibrils. The disruption of the fibrils increases the accessibility of the cellulose to the enzymes during hydrolysis.

Steam explosion is typically initiated at a temperature of 160–260°C (corresponding pressure 0.69–4.83 MPa) for several seconds to a few minutes before the material is exposed to atmospheric pressure. The process causes hemicellulose degradation and lignin transformation due to high temperature, thus increasing the potential of cellulose hydrolysis (Ballesteros et al.2006, Chornet et al. 1988, Focher et al. 1991).

However, some disadvantages have been seen when using this process. Dilute acids are required to be added during softwood pretreatment or even when increased yields are warranted for lower acetylated feedstock. The factors that affect steam explosion pretreatment are residence time, temperature, chip size and moisture content (Duff and Murray 1996). Recent studies indicate that lower temperature and longer residence time are more favorable (Wright 1998).

Liquid Hot Water (LHW)

Much like the steam-explosion process, liquid hot water (LHW) pretreatment uses water at elevated temperatures and high pressures to maintain its liquid form in order to promote disintegration and separation of the lignocellulosic matrix. Temperatures can range from 160°C to 240°C over lengths of time ranging from a few minutes up to an hour with temperatures dominating the types of sugar formation and time dominating the amount of sugar formation (Yu et al. 2010).

This process has been found to be advantageous from a cost standpoint in that no additives such as acid catalysts are required. Furthermore, expensive reactor systems have not been necessary to use due to the low corrosive nature of this pretreatment technique. Neutralization of degradation products is not needed due to their fractionation and utilization in the liquid fraction. In the same sense, inhibitory products have not been reported to form overwhelmingly in the respective fractions allowing higher yields under specific conditions.

Ammonia Fiber Explosion (AFEX)

The ammonia fiber/freeze explosion (AFEX) process is another physicochemical process, much like steam explosion pretreatment, in which the biomass material is subjected to liquid anhydrous ammonia under high pressures and moderate temperatures and is then rapidly depressurized. The moderate temperatures (60°C to 100°C) are significantly less than that of the steam explosion process, thus allowing less energy input and overall cost reduction associated with the process (Alizadeh et al. 2005, Teymouri et al. 2004, Chundawat et al. 2007).

There have been extensive literature reviews on this type of pretreatment over the last decade, focusing on the advantages and disadvantages of the AFEX process used for different feedstocks (Sun and Cheng 2002, Mosier 2005). An overview of some of the advantages include lower moisture content, lower formation of sugar degradation products due to moderate conditions, 100% recovery of solid material, and the ability for ammonia to lessen lignin's effect on enzymatic hydrolysis. A smaller number of disadvantages can be seen in the form of higher costs due to recycle and treatment of chemicals that are being used.

Ammonia Recycle Percolation (ARP)

Ammonia recycle percolation (ARP) has been paired with the AFEX pretreatment process by many authors, but it can have some different characteristics that need to be taken into consideration when looking at different pretreatment options (Kim and Lee 2005). In this process, aqueous ammonia of concentration between 5-15% (wt%) is sent through a packed bed reactor containing the biomass feedstock at a rate of about 5 ml/min. The advantage with this process over AFEX is its ability to remove a majority of the lignin (75–85%) and solubilize more than half of the hemicellulose (50–60%) while maintaining high cellulose content (Kim and Lee 2005). Primarily, herbaceous biomass has been most treated with this process: 60-80% delignification has been achieved for corn stover and 65–85% delignification for switchgrass (Iyer et al. 1996).

Supercritical Fluid (SCF) Pretreatment

A supercritical fluid is a material which can be either liquid or gas, used in a state above the critical temperature and critical pressure where gases and liquids can coexist. It shows unique properties that are different from those of either gases or liquids under standard conditions-it possesses a liquid like density and exhibits gas-like transport properties of diffusivity and viscosity (King and Srinivas 2009). Thus, SCF has the ability to penetrate the crystalline structure of lignocellulosic biomass overcoming the mass transfer limitations encountered in other pretreatments. The lower temperatures used in the process aids in the stability of the sugars and prevents degradation observed in other pretreatments. Kim and Hong 2001 investigated supercritical CO_2 pretreatment of hardwood (Aspen) and southern yellow pine with varying moisture contents followed by enzymatic hydrolysis. SCF pretreatment showed significant enhancements in sugar yields when compared to thermal pretreatments without supercritical CO_2. Alinia and coworkers (2010) investigated the effect of pretreatment of dry and wet wheat straw by supercritical CO_2 alone and by a combination of CO_2 and steam under different operating conditions (temperature and residence time in the reactors). It was found that a combination of supercritical CO_2 and steam gave the best overall yield of sugars.

Biological Pretreatment Methods

Biological pretreatment uses microorganisms and their enzymes selectively for delignification of lignocellulosic residues and has the advantages of a low-energy demand, minimal waste production and a lack of environmental effects. In biological pretreatment processes, microorganisms such as brown-, white- and soft-rot fungi are used to degrade lignin and hemicellulose in waste materials Schurz (1978). White-rot basidiomycetes possess the capabilities to attack lignin. *Penicillium chrysosporium*, for example, has been shown to non-selectively attack lignin and carbohydrate (Anderson and Akin 2008). *P. chrysosporium* has been successfully used for biological pretreatment of cotton stalks by solid state cultivation (SSC) and results have shown that the fungus facilitates the conversion into ethanol (Shi et al., 2008). Brown rots mainly attack cellulose, while white and soft rots attack both cellulose and lignin. White-rot fungi are the most effective basidiomycetes for biological pretreatment of lignocellulosic materials (Fan et al. 1987). Other basidiomycetes such as *Phlebia radiata, P. floridensis* and *Daedalea flavida*, selectively degrade lignin in wheat straw and are good choices for delignification of lignocellulosic residues (Arora and Chander 2002). *Ceriporiopsis subvermispora*, however, lacks cellulases (cellobiohydrolase activity) but produces manganese peroxide and laccase, and selectively delignifies several different wood species (Ferraz 2003). The advantages of biological pretreatment include low energy requirement and mild environmental conditions. However, the rate of hydrolysis in most biological pretreatment processes is very low.

Hydrolysis of Pretreated Biomass

After pretreatment, the released cellulose and hemicelluloses are hydrolyzed to soluble monomeric sugars (hexoses and pentoses) using cellulases and hemicellulases, respectively. The initial conversion of biomass into sugars is a key bottleneck in the process of biofuel production and new biotechnological solutions are needed to improve their efficiency, which would lower the overall cost of bioethanol production. Enzymatic hydrolysis has been considered key to cost-effective bioethanol in the long run, and the reaction is carried out with mainly cellulase and hemicellulase for cellulose and hemicellulose, respectively. The advantages of using enzyme (cellulase) over acid is to eliminate corrosion problems and lower maintenance costs with mild processing conditions to give high yields.

Despite the fact that some fungal strains have the advantages of being thermostable and producing cellulases, most of these fungal strains do not produce sufficient amounts of one or more lignocellulolytic enzymes required for efficient bioconversion of lignocellulosic residues to fermentable sugars. In addition, plant cell walls are naturally resistant to microbial and enzymatic (fungal and bacterial) deconstruction, collectively known as 'biomass recalcitrance' (Himmel et al.2007). These rate-limiting steps in the bioconversion of lignocellulosic residues to ethanol remain one of the most significant hurdles to producing economically feasible cellulosic ethanol. Improving fungal hydrolytic activity and finding stable enzymes capable of tolerating extreme conditions has become a priority in many recent studies.

Fungal Extracellular Cellulases

Enzymatic saccharification of lignocellulosic materials such as sugarcane bagasse, corncob, rice straw, *Prosopis juliflora, Lantana camara,* switch grass, saw dust, and forest residues by cellulases for biofuel production is perhaps the most popular application currently being investigated (Kuhad et al. 2010, Sukumaran et al. 2005). Both bacteria and fungi can produce glucanases (cellulases) that hydrolyze of lignocellulosic materials. These microorganisms can be aerobic or anaerobic and mesophilic or thermophilic. Bacteria belonging to genera of *Clostridium, Cellulomonas, Bacillus,Thermomonospora, Ruminococcus, Bacteriodes, Erwinia, Acetovibrio, Microbispora,* and*Streptomyces* are known to produce cellulase (Bisaria 1998). Anaerobic bacterial species such as*Clostridium phytofermentans, Clostridium thermocellum, Clostridium hungatei, and Clostridium papyrosolvens* produces cellulases with high specific activity (Duff and Murray 1996, Bisaria 1998). Most commercial glucanases (cellulases) are produced by *Trichoderma ressei* and β-D-glucosidase is produced from *Aspergillus niger* (Kaur et al. 2007). Fungi known to produce cellulases include*Sclerotium rolfsii, Phanerochaete chrysosporium* and various species of *Trichoderma, Aspergillus,Schizophyllum and Penicillium* (Sternberg 1976, Fan et al. 1987, Duff and Murray 1996). Among the fungi, *Trichoderma* species have been extensively studied for cellulase production (Sternberg1976).

High temperature and low pH tolerant enzymes are preferred for the hydrolysis due to the fact that most current pretreatment strategies rely on acid and heat (Turner et al. 2007). In addition, thermostable enzymes have several advantages including higher specific activity and higher stability which improve the overall hydrolytic performance (Viikari et al. 2007). Ultimately, improvement in catalytic efficiencies of enzymes will reduce the cost of hydrolysis by enabling lower enzyme dosages. Some fungal strains such as *T. emersonii* (Grassick et al. 2004),*Chaetomium thermophilum* (Li et al. 2006) and *Corynascus thermophilus* (Rosgaard et al. 2006) can produce thermostable enzymes which are stable and active at elevated temperatures (60°C) well above their optimum growth temperature (30-55°C) (Maheshwari et al., 2000). Due to the promising thermostability and acidic tolerance of thermophilic fungal enzymes, they have good potential to be used for hydrolysis of lignocellulosic residues at industrial scales.

The anaerobic bacteria *Clostridium thermocellum* and *Clostridium cellulovorans* and the filamentous fungus *Trichoderma reesei* are well known as strongly cellulolytic and xylanolytic microorganisms. *C. thermocellum* and *C. cellulovorans* produce a cellulosome complex consisting of cellulase and hemicellulase organized on the cell surface (Demain et al. 2005); *T. reesei*, meanwhile, extracellularly secretes three types of cellulolytic enzyme, including five endoglucanases (EG [EC 3.2.1.4]) (Pere et al. 2001, Dienes et al. 2004), two cellobiohydrolases (CBH [EC 3.2.1.91]) (Bayer et al. 1998), and two β-glucosidases (BGL [EC 3.2.1.21]) (Sang-Mok and Koo 2001). Endoglucanases act randomly against the amorphous region of the cellulose chain to produce reducing and nonreducing ends for cellobiohydrolases, which produce cellobiose from reducing or nonreducing ends of crystalline cellulose. Cellulose chains are thus efficiently degraded to soluble cellobiose and cellooligosaccharides by the endo-exo synergism of EG and CBH (Hebeish and Ibrahim 2007). In the last step of enzymatic cellulose degradation, cellooligosaccharides are hydrolyzed to glucose by β-glucosidase. In addition to endo-exo synergism, exo-exo synergism between two cellobiohydrolases has also been reported.

Fungal Hemicellulases

Several different enzymes are needed to hydrolyze hemicelluloses, due to their heterogeneity (Saha 2003). Xylan is the most abundant component of hemicellulose contributing over 70% of its structure. Xylanases are able to hydrolyze β-1,4 linkages in xylan and produce oligomers which can be further hydrolyzed into xylose by β-xylosidase. Not surprisingly, additional enzymes such as β-mannanases, arabinofuranosidases or α-L-arabinases are needed depending on the hemicellulose composition which can be mannan-based or arabinofuranosyl-containing. Also similarly to cellulases, most of the hemicellulases are glycoside hydrolases (GHs), although some hemicellulases belong to carbohydrate esterases (CEs) which hydrolyze ester linkages of acetate or ferulic acid side groups (Shallom and Shoham 2003). A mixture of hemicellulases or pectinases with cellulases exhibited a significant increase in the extent of cellulose conversion (Ghose and Bisaria 1979, Beldman et al. 1984). Many fungal species such as *Trichoderma, Penicillium,Aspergillus* and *T. emersonii* have been reported to produce large amounts of extracellular cellulases and hemicellulases.

Fungal Ligninases

Fungi degrade lignin by secreting enzymes collectively termed "ligninases". These include two ligninolytic families; i) phenol oxidase (laccase) and ii) peroxidases [lignin peroxidase (LiP) and manganese peroxidase (MnP)] (Martinez et al. 2005). White-rot basidiomycetes such as *Coriolus versicolor* (Wang et al. 2003), *P. chrysosporium* and *T. versicolor* (Moredo et al. 2003) have been found to be the most efficient lignin-degrading microorganisms studied. Interestingly, LiP is able to oxidize the non-phenolic part of lignin, but it was not detected in many lignin degrading fungi. In addition, it has been widely accepted that the oxidative ligninolytic enzymes are not able to penetrate the cell walls due to their size. Thus, it has been suggested that prior to the enzymatic attack, low-molecular weight diffusible reactive oxidative compounds have to initiate changes to the lignin structure and hemicellulose, fungal cellulosomes are much less well characterized compared to bacterial cellulosomes.

Fermentation

In the fermentation process, the hydrolytic products including monomeric hexoses (glucose, mannose and galactose) and pentoses (xylose and arabinose) will be fermented to valuable products such as ethanol. Among these hydrolytic products, glucose is normally the most abundant, followed by xylose or mannose and other lower concentration sugars.

The last two steps of bioconversion of pretreated lignocellulolytic residues to ethanol (hydrolysis and fermentation) can be performed separately (SHF) or simultaneously (SSF). In the separate hydrolysis and fermentation (SHF), the hydrolysate products will be fermented to ethanol in a separate process. The advantage of this method is that both processes can be optimized individually (e.g. optimal temperature is 45-50°C for hydrolysis, whereas it is 30°C for fermentation). However, its main drawback is the accumulation of enzyme-inhibiting end-products (cellobiose and glucose) during the hydrolysis. This makes the process inefficient, and the costly addition of β-glucosidase is needed to overcome end-product inhibition (Elumalia and Thangavelu2010).

Further process integration can be achieved by a process known as consolidated bioprocessing (CBP) which aims to minimize all bioconversion steps into one step in a single reactor using one or more microorganisms. CBP operation featuring cellulase production, cellulose/hemicellulose hydrolysis and fermentation of 5- and 6- carbon sugars in one step have shown the potential to provide the lowest cost for biological conversion of cellulosic biomass to fuels, when processes relying on hydrolysis by enzymes and/or microorganisms are used (Lynd et al. 2005).

The simultaneous saccharification and fermentation (SSF) process was first studied by Takagi et al. (1977) for cellulose conversion to ethanol. The SSF process was originally developed for lignocellulosic biomass by researchers at Gulf Oil Company in 1974 (Blotkamo et al. 1978). The SSF process eliminates expensive equipment and reduces the probability of contamination by unwanted organisms that are less ethanol tolerant than the microbes selected for fermentation (Szczodrak 1989).

SSF combines the enzymatic saccharification of polymeric cellulose to simple monomeric forms such as glucose and its eventual fermentation by yeast to ethanol in the same vessel (Ikwebe and Harvey 2011). In simultaneous saccharification and fermentation (SSF), however, the end-products will be directly converted to ethanol by the microorganism. Therefore, addition of high amounts of β-glucosidase is not necessary and this reduces the ethanol production costs (Stenberg et al. 2000). Rapid conversion of the glucose into ethanol by yeast results in faster rates, higher yields, and greater ethanol concentrations than possible for SHF. The presence of ethanol in the fermentation broth also makes the mixture less vulnerable to invasion by unwanted microorganisms (Sasikumar and Viruthagiri 2010). However, the main drawback of SSF is the need to compromise processing conditions such that temperature and pH are suboptimal for each individual step. The development of recombinant yeast strains with improved thermotolerance can enhance the performance of SSF (Galbe and Zacchi 2002). It is reported that the major inefficiencies of biochemical process for lignocellulosic bioethanol production were identified as the simultaneous saccharification and fermentation (SSF) process accounting for 27% of the lost energy by thermodynamic analysis (Sohel and Jack 2010). Alkasrawi et al. (2003) reported that addition of surfactants as an additive in SSF can significantly lower the operational cost of the process because it increases the conversion rate of cellulose to glucose. Addition of Tween-20, 2.5 g/l not only reduces the time required to attain maximum ethanol concentration, but also enhances enzyme activity in the liquid fraction at the end of SSF, probably by preventing unproductive binding of the cellulases to lignin, which could facilitate enzyme recovery.

Over the years, various groups have worked on the SSF process to improve the choice of enzymes, fermentative microbes, biomass pretreatment, and process conditions. Extensive studies on SSF have since been conducted focusing on the production of ethanol from cellulosic substrates. Phillipidis et al. (1993) have studied the enzymic hydrolysis of cellulose in an attempt to optimize SSF performance. Ghose et al. (1984) have increased ethanol productivity by employing a vacuum cycling in an SSF process using lignocellulosic substances. Zhu et al. (2006b) evaluated the suitability of production of ethanol from the microwave-assisted alkali pretreated wheat straw, the simultaneous saccharification and fermentation (SSF) of the microwave-assisted and conventional alkali pretreated wheat straw to ethanol.

Candida brassicae is accepted as the yeast of choice as far as SSF is considered, although both*Saccharomyces cerevisiae* and *S. carlsbergensis* have been found to offer similar rates. Several other yeasts as well as the bacteria *Zymomonas mobilis* have been studied with cellulose from *T.ressei* mutants for SSF processes. Researchers have also examined several combinations of enzymes with *Z. mobilis*, *S. cerevisiae*, and other ethanol producer, but they have only considered substrate levels lower than necessary to prove economic viability. Wyman et al. (1986) evaluated the cellobiose-fermenting yeast *Brettanomyces clausenii* for the SSF of cellulose to ethanol.

There are number of different methods to quantitate ethanol in samples. HPLC has been utilized to monitor the fermentation process This method has the advantage of being able to monitor not only the production of ethanol, but also the reaction substrates and byproducts (Hall and Reuter 2007). Fourier transform infrared spectroscopy (Sharma et al. 2009), gas chromatography (Wang et al.2003), and Infrared (Lachenmeier et al. 2010) technologies have also been used to detect and quantitate ethanol in samples. While FTIR requires a large investment in instrumentation, the use or less expensive IR technology has been demonstrated to be just as accurate (Lachenmeier et al.2010). Gerchman et al. (2012) developed a cheap and rapid approach for ethanol quantification in aqueous media during fermentation steps as part of the conversion of biomass to ethanol. The suggested method requires a sample of a small volume and consists of organic extraction, followed by direct use of gas chromatography with a flame ionization detector (GC-FID). The feasibility of such approach is obvious since there is no need for the head-space system, distillation, expensive reagents and sophisticated equipment. The proposed method was also tested for its 'real-life' applicability for ethanol quantification from fermentation process.

According to Keller and Bryan (2000), distillation is still a "formidable competitor" as a major separation method even though much research has been thrust on its alternatives. Hence, distillation, especially simple distillation, tends to be the first choice in industry for separating a liquid mixture; other methods, including complex distillation, e.g., azeotropic distillation, come into play only when simple distillation is deemed to be technically infeasible or economical inviable because of typically three large stainless steel distillation towers, stainless steel heat exchangers and price of stainless up 400% in last six years, high operating costs because 280 MMBTU/hr energy is consumed (100 MGPY ethanol). Mole sieve drying adds to energy costs and that's why energy costs up significantly with price of crude oil.

Under certain circumstances, retrofitting of an existing process can be economically far more viable than constructing a new process, especially when the financial resources are limited and/or when short term needs are to be met under a tight time constraint. Developing economically viable fermentation processes requires efficient downstream processing: selective product removal and avoiding byproduct streams. "ESepis a modular, low-energy process for the recovery of ethanol from fermentation broth with an estimated reduction of up to 60% in both capital and operating costs versus conventional distillation. Use of non-stainless steel components also results in a substantial reduction in construction time". It is applicable to new ethanol plants (corn, sugar and cellulosic). It replaces whole distillation train and mole sieve dryer. With new plants it reduces overall energy consumption by >60% (ESep 2008).

The utilization of pervaporation for the production of absolute (anhydrous) ethanol through its coupling with the previous distillation step has been reported. The modeling and optimization of the process using MINLP tools showed 12% savings in the production costs considering a 32% increase in membrane area and the reduction in both reflux ratio and ethanol concentration in the distillate of the column (Lelkes et al. 2000, Szitkai et al. 2002). Through pilot-plant studies, the integration of distillation process with the pervaporation has been achieved resulting in good indexes in terms of energy savings. These savings are due to the low operation costs of pervaporation and to the high yield of dehydrated ethanol, typical of pervaporation processes (Tsuyomoto et al. 1997). The comparison between azeotropic distillation using benzene and pervaporation system using multiple membrane modules showed that, at the same ethanol production rate and quality (99.8 wt.%), operation costs, including the membrane replacement every 2–4 years, are approximately 1/3–1/4 of those of azeotropic distillation.

Methods Used to Improve Fungal Enzyme Production, Activity and/or Stability

In order to produce ethanol industrially, the fermentative microorganism needs to be robust. The utilization of all the sugars generated from lignocellulosic hydrolysate is essential for the economical production of ethanol (Saha 2003). The conventional ethanol fermenting yeast (*Saccharomyces cerevisiae*) or bacterium (*Zymomonas mobilis*) cannot ferment multiple sugar substrates to ethanol (Bothast et al. 1999). A major technical hurdle to converting lignocellulose to ethanol is developing an appropriate microorganism for the fermentation of a mixture of sugars such as glucose, xylose, arabinose, and galactose (Bothast et al. 1999). A number of recombinant microorganisms such as *Escherichia coli, Klebsiella oxytoca, Z. mobilis,* and *S. cerevisiae* have been developed over the last 25 years with a goal of fermenting mixed sugars to ethanol (Alterthum and Ingram 1989, Ohta et al. 1991, Zhang et al. 1995, Ho et al. 1998). Saha and Cotta's (2011) research unit has developed a recombinant *E. coli* (strain FBR5) that can ferment mixed multiple sugars to ethanol (Dien et al. 2000). The strain carries the plasmid pLOI297, which contains the genes for pyruvate decarboxylase (pdc) and alcohol dehydrogenase (adh) from *Z.mobilis* necessary for efficiently converting pyruvate into ethanol (Alterthum and Ingram 1989).

Technologies required for bioconversion of lignocelluloses to ethanol and other valuable products are currently available but need to be developed further in order to make biofuels cost competitive compared to other available energy resources such as fossil fuels. The most recent and important improvements in production/activity of fungal enzymes using different techniques such as mutagenesis, co-culturing and heterologous gene expression of cellulases are discussed below.

Mutagenesis

Many fungal strains have been subjected to extensive mutagenesis studies due to their ability to secrete large amounts of cellulose-degrading enzymes. Cellulolytic activity of *T. reesei* QM6a has been improved by using different mutagenesis techniques including UV-light and chemicals, resulting in the mutant QM 9414 with higher filter paper activity (FPA) (Mandels et al. 1971). *T.reesei* RUT-C30 is one of the best known mutants, producing 4–5 times more cellulase than the wild-type strain (QM 6a). A recent study by Kovacs et al. 2008 has shown that wild-type*Trichoderma atroviride* (F-1505) produces the most cellulase among 150 wild-type *Trichoderma*. Moreover, *T. atroviride* mutants were created by mutagenesis using *N*-methyl-*N'*-nitro-*N*-nitrosoguanidine (NTG) as well as UV-light. These *T. atroviride* mutants (e.g. *T. atroviride* TUB F-1724) produce high levels of extracellular cellulases as well as β-glucosidase when they are grown on pretreated willow. Cellulase and xylanase activities in *Penicillium verruculosum* 28 K mutants were improved about 3-fold using four cycles of UV mutagenesis. The enzyme production was further improved by 2- to 3-fold in a two-stage fermentation process using wheat bran, yeast extract medium and microcrystalline cellulose as the inducer (Soloveva et al. 2005).

Site-directed mutagenesis (SDM) plays a central role in the characterization and improvement of cellulases including their putative catalytic and binding residues. The application of SDM it was found that Glu 116 and 200 are the catalytic nucleophile and acid–base residues in *Hypocrea jecorina* (anamorph *T. reesei*) Cel12A, respectively. In the study, mutant enzymes were produced where Glu was replaced by Asp or Gln at each position (E116D/Q and E200D/Q). The specific activity of these mutants was reduced by more than 98%, suggesting the critical role of these two residues in the catalytic function of the enzyme (Okada et al. 2000). In another study, the thermostable endo-1,4-β-xylanase (XynII) mutants from *T. reesei* were further mutated to resist inactivation at high pH by using SDM. All mutants were resistant to thermal inactivation at alkaline pH. For example, thermotolerance for one mutant (P9) at pH 9 was increased approximately 4–5°C, resulting in better activity in sulphate pulp bleaching compared to the reference (Fenel et al.2006). Also, the catalytic efficiency and optimum pH of *T. reesei* endo-β-1, 4-glucanase II were improved by saturation mutagenesis followed by random mutagenesis and two rounds of DNA shuffling. The pH optimum of the variant (Q139R/L218H/W276R/N342T) was shifted from 4.8 to 6.2, while the enzyme activity was improved more than 4.5-fold (Qin et al. 2008). Moreover, the stability of *T. reesei* endo-1, 4-β-xylanases II (XynII) was increased by engineering a disulfide bridge at its N-terminal region. In fact, two amino acids (Thr-

2 and Thr-28) in the enzyme were substituted by cysteine (T2C:T28C mutant) resulting in a 15°C increase in thermostability (Fenel et al. 2004).

Co-culturing

Fungal co-culturing offers a means to improve hydrolysis of lignocellulosic residues, and also enhances product utilization which minimizes the need for additional enzymes in the bioconversion process. In the case of cellulose degradation, for example, all three enzymatic components (EG, CBH and β-glucosidase) have to be present in large amounts. However, none of the fungal strains, including the best mutants, are able to produce high levels of the enzymes at the same time. *T.reesei* for example produces CBH and EG in high quantities whereas its β-glucosidase activity is low (Stockton et al. 1991). *A. niger* however, produces large amounts of β-glucosidase, but has limited EG components (Kumar et al. 2008). In addition, hemicellulose hydrolysis must also be considered when lignocellulosic residues are subjected to biomass conversion. However, this will be determined by the pretreatment methods. Specifically in an alkali pretreatment method, a part of lignin will be removed and thus hemicellulose has to be degraded by the use of hemicellulases, whereas in acid-catalyzed pretreatment, the hemicellulose layer will be hydrolyzed (Hahn et al.2006). Again, some fungal strains have been shown to work more efficiently on cellulosic residues whereas others produce more hemicellulolytic enzymes and efficiently hydrolyze hemicellulosic portions (Howard et al. 2003). Conversion of both cellulosic and hemicellulosic hydrolytic products in a single process can be achieved by co-culturing two or more compatible microorganisms with the ability to utilize the materials. In fact, in nature, lignocellulosic residues are degraded by multiple co-existing lignocellulolytic microorganisms.

Mixed fungal cultures have many advantages compared to their monocultures, including improving productivity, adaptability and substrate utilization. Improving fungal cellulolytic activity of *T. reesei*and *A. niger* by co-culturing was the subject of extensive research including studies done by Maheshwari et al. (1994), Ahamed et al. (2008) and Juhasz et al. 2003. Moreover, other fungal strains have been co-cultured to obtain better cellulolytic activity such as co-culturing of *T. reesei*RUT-C30 and *A. phoenicis* (Duff 1985) or *A. ellipticus* and *A.*

fumigatus (Gupte and Madamwar1997). There are a few examples of co-culturing fungal strains for the purpose of combining cellulose and hemicellulose hydrolysis such as co-culturing *T. reesei* D1-6 and *A. wentii* Pt 2804 in a mixed submerged culture (Panda et al. 1983) or co-culturing *T. reesei* LM-UC4 and *A. phoenicis*QM329 using ammonia-treated bagasse (Duenas et al. 1995). In the both cases, enzyme activity for cellulases and hemicellulases was significantly increased. The main drawback of co-culturing however is the complexity of growing multiple microorganisms in the same culture (Lynd et al.2002).

Metabolic Engineering

Metabolic engineering is a powerful method to improve, redirect, or generate new metabolic reactions or whole pathways in microorganisms. This enables one microorganism to complete an entire task from beginning to end. This can be done by altering metabolic flux by blocking undesirable pathway(s) and/or enhancement of desirable pathway(s). For example by application of homologous recombination, the production of *T. reesei* β-glucosidase I was enhanced using xylanase (*xyn3*) and cellulase (*egl3*) promoters which improved β-glucosidase activity to 4.0 and 7.5 fold compared to the parent, respectively. This will permit one fungal strain such as *T. reesei*to be more efficient on hydrolysis of cellulose to glucose which improve the yield and therefore lower the cost (Rahman et al. 2009). Becker and Boles (2003) described the engineering of a*Saccharomyces cerevisiae* strain able to utilize the pentose sugar L-arabinose for growth and to ferment it to ethanol. Expanding the substrate fermentation range of *S. cerevisiae* to include pentoses is important for the utilization of this yeast in economically feasible biomass-to-ethanol fermentation processes. After overexpression of a bacterial L-arabinose utilization pathway consisting of *Bacillus subtilis* AraA and *Escherichia coli* AraB and AraD and simultaneous overexpression of the L-arabinose-transporting yeast galactose permease, we were able to select an L-arabinose-utilizing yeast strain by sequential transfer in L-arabinose media. High L-arabinose uptake rates and enhanced transaldolase activities favor utilization of L-arabinose.

Shaw et al. 2008 engineered *Thermoanaerobacterium saccharolyticum*, a thermophilic anaerobic bacterium that ferments xylan and biomass-derived sugars, to produce ethanol at high yield.

Knockout of genes involved in organic acid formation (acetate kinase, phosphate acetyltransferase, and L-lactate dehydrogenase) resulted in a strain able to produce ethanol as the only detectable organic product and substantial changes in electron flow relative to the wild type. Glucose and xylose are co-utilized and utilization of mannose and arabinose commences before glucose and xylose are exhausted.

Heterologous Expression

Heterologous expression is a powerful technique to improve production yield of enzymes, as well as activity. In order to make a robust lignocellulolytic fungal strain, many different fungal cellulases with higher and/or specific activity based on the need for a functional cellulase system in the organism have been cloned and expressed. For example, thermostable β-glucosidase (*cel3a*) from thermophilic fungus *T. emersonii* was expressed in *T. reesei* RUT-C30 using a strong *T.reesei cbh1* promoter. The expressed enzyme has been shown to be highly thermostable (optimum temperature at 71.5°C) with high specific activity (Murray et al. 2004). In the study for the improvement of biofinishing of cotton, *T. reesei* cellobiohydrolase (I & II) were overexpressed using additional copy(s) of the genes cloned under *T. reesei cbh1* promoter. The results have shown that the expression of CBHI was increased to 1.3- and 1.5-fold with one or two additional copies of the gene, respectively.

Immobilization

Immobilization of microbial cells and enzymes have showed certain technical and economical advantages over free cell system. Using immobilized enzymes not only leads to greater product purity, cleaner processes, and economic operational costs but also makes the use enzyme cost effective and recoverable (Meena and Raja 2004). The immobilized biocatalysts have been extensively investigated during last few decades. An immobilized cellobiase enzyme system has been used in the enzymatic hydrolysis of biomass for the generation of cellulosic ethanol (Das et al., 2011). Production of alcohol and biodiesel fuel from triglycerides using immobilized lipase has been carried out using porous kaolinite particle as a carrier (Iso et al. 2001).

The use of an immobilized yeast cell system for alcoholic fermentation is an attractive and rapidly expanding research area because of its additional technical and economical advantages compared with the free cell system. A reduction in the ethanol concentration in the immediate microenvironment of the organism due to the formation of a protective layer or specific adsorption of ethanol by the support may act to minimize end product inhibition. The most significant advantages of immobilized yeast cell systems are the ability to operate with high productivity at dilution rates exceeding the maximum specific growth rate, the increase of ethanol yield and cellular stability and the decrease of process expenses due to the cell recovery and reutilization (Lin and Tanaka 2006). Other advantages of immobilized cell system over presently accepted batch or continuous fermentations with free-cells are: greater volumetric productivity as a result of higher cell density; tolerance to higher concentrations of substrate and products; lacking of inhibition; relative easiness of downstream processing etc. in different types of bioreactors, such as packed bed reactor, fluidized bed reactor, gaslift reactor and reactor with magnetic field (Ivanova et al. 1996, Sakai et al. 1994; Perez et al. 2007).

Perspective techniques for yeasts immobilization can be divided into four categories: attachment or adsorption to solid surfaces (wood chips, delignified brewer's spent grains, DEAE cellulose, and porous glass), entrapment within a porous matrix (calcium alginate, k-carrageenan, polyvinyl alcohol, agar, gelatine, chitosan, and polyacrilamide), mechanical retention behind a barrier (microporous membrane filters, and microcapsules) and self-aggregation of the cells by flocculation (Ivanova et al. 2011).

Process Integration

One of the most important approaches for the design of more intensive and cost-effective process configurations is process integration. Process integration looks for the integration of all operations involved in the production of fuel ethanol. This can be achieved through the development of integrated bioprocesses that combine different steps into one single unit. Thus, reaction–separation integration by removing ethanol from the zone where the biotransformation takes place, offers several opportunities for increasing product yield and consequently reducing product costs. Other forms of integration may significantly decrease energetic costs of specific flowsheet configurations for ethanol production. Process integration is gaining more and more interest due to the advantages related to its application in the case of ethanol production: reduction of energy costs, decrease in the size and number of process units, intensification of the biological and downstream processes. Integration of fermentation and separation processes for reduction of product inhibition, development of efficient cogeneration technologies using cane bagasse, development of CBP, application of membrane technology (e.g. for ethanol removal or dehydration) are examples of process integration.

CONCLUSIONS

Lignocellulolytic microorganisms, especially fungi, have attracted a great deal of interest as biomass degraders for large-scale applications due to their ability to produce large amounts of extracellular lignocellulolytic enzymes. Many successful attempts have been made to improve fungal lignocellulolytic activity including recombinant and non-recombinant techniques. Process integration has also been considered for the purpose of decreasing the production cost, which was partly achieved by performing hydrolysis and fermentation in a single reactor (SSF) using one or more microorganisms (co-culturing).

These laboratory improvements should now be verified in pilot and demonstration plants. Scaling up the production of lignocellulosic ethanol, however, requires further reduction of the production cost. Thus, in order to improve the technology and reduce the production cost, two major issues have to be addressed: i) improving technologies to overcome the recalcitrance of cellulosic biomass conversion (pretreatment, hydrolysis and fermentation) and ii) sustainable production of biomass in very large amounts.

Future Prospects

It is considered that lignocellulosic waste will become the main feedstock for ethanol production in the near future. In the case of large scale biomass production, additional waste stocks can be tested and used as substrates to meet the needs. On the other hand, biotechnological approaches including systems biology and computational tools are likely good candidates to overcome these issues. Future trends for costs reduction should include more efficient pretreatment of biomass, improvement of specific activity and productivity of cellulases, improvement of recombinant microorganisms for a greater assimilation of all the sugars released during the pretreatment and hydrolysis processes, and further development of co-generation system. Undoubtedly, ongoing research on genetic and metabolic engineering will make possible the development of effective and stable strains of microorganisms for converting cellulosic biomass into ethanol. Process engineering will play a central role for the generation, design, analysis and implementation of technologies improving the indexes of global process, or for the retrofitting of employed bioprocesses. Undoubtedly, process intensification through integration of different phenomena and unit operations as well as the implementation of consolidated bioprocessing of different feedstocks into ethanol (that requires the development of tailored recombinant microorganisms), will offer the most significant outcomes during the search of the efficiency in fuel ethanol production. This fact will surely imply a qualitative improvement in the industrial production of fuel ethanol in the future.

ACKNOWLEDGEMENTS

The authors are thankful to Chancellor, ITM University, Gwalior, MP, and Dr. J.L. Bhat, Dean, Life Sciences, ITM University, Gwalior, MP, for their constant support and encouragement.

REFERENCES

1. Ahamed A, Vermette P (2008) Enhanced enzyme production from mixed cultures of Trichoderma reesei RUT-C30 and Aspergillus niger LMA grown as fed batch in a stirred tank bioreactor. Biochem Eng J 42:41-46
2. Alinia R, Zabihi S, Esmaeilzadeh F, Kalajahi JF (2010) Pretreatment of wheat straw by supercritical CO2 and its enzymatic hydrolysis for sugar production. Biosystems Eng 107(1):61-66

3. Alizadeh H, Teymouri F, Gilbert TI, Dale BE (2005) Pretreatment of switchgrass by ammonia fiber explosion (AFEX). Appl Biochem Biotechnol 124(1–3):1133-1141
4. Alkasrawi M, Eriksson T, Borjesson J, Wingren A, Galbe M, Tjerneld F, Zacchi G (2003) The effect of tween-20 on simultaneous saccharification and fermentation of softwood to ethanol. Enzyme and Microbial Technol 33:71-78
5. Alterthum F, Ingram LO (1989) Efficient ethanol production from glucose, lactose, and xylose by recombinant Escherichia coli. Appl Environ Microbiol 55:1943-1948
6. Anderson WF, Akin DE (2008) Structural and chemical properties of grass lignocelluloses related to conversion for biofuels. J Ind Microbiol Biotechnol 35:355-366
7. Arora DS, Chander MKGP (2002) Involvement of lignin peroxidase, manganese peroxidase and laccase in degradation and selective ligninolysis of wheat straw. Int Bioterior Biodegrad 50:115-120
8. Aswathy US, Sukumaran RK, Lalitha D, Rajeshree KP, Singhania RR, Pandey A (2010) Bio-ethanol from water hyacinth biomass: an evaluation of enzymatic saccharification strategy. Bioresour Technol 101:925-930
9. Azzam M (1989) Pretreatment of cane bagasse with alkaline hydrogen peroxide for enzymatic hydrolysis of cellulose and ethanol fermentation. J Environ Sci Health B 24(4):421-433
10. Balan V, Sousa LDC, Chundawat SPS, Marshall D, Sharma LN, Chambliss CK, Dale BE (2009) Enzymatic digestibility and pretreatment degradation products of AFEX-treated hardwoods (Populusnigra). Biotechnol Prog 25:365-375
11. Ballesteros I, Negro MAJ, Oliva JM, Cabanas A, Manzanares P, Ballesteros M (2006) Ethanol production from steam-explosion pretreated wheat straw. Appl Biochem Biotechnol 130(1–3):496-508
12. Ballesteros M, Oliva JM, Negro MJ, Manzanares P, Ballesteros I (2004) Ethanol from lignocellulosic materials by a simultaneous saccharification and fermentation process (SFS) with Kluyveromyces marxianus CECT 10875. Process Biochem 39:1843-1848

13. Banerjee S, Sen R, Pandey RA (2009) Evaluation of wet air oxidation as a pretreatment strategy for bioethanol production from rice husk and process optimization. Biomass Bioenergy 33(12):1680-1686

14. Bayer EA, Chanzy H, Lamed R, Shoham Y (1998) Cellulose, cellulases and cellulosomes. Curr Opin Struct Biol 8(5):548-557

15. Becker J, Boles E (2003) A modified Saccharomyces cerevisiae strain that consumes l-arabinose and produces ethanol. Appl Environ Microbiol 69(7):4144-4150

16. Beldman G, Rombouts FM, Voragen AGJ, Pilnik W (1984) Application of cellulase and pectinase from fungal origin for the liquefaction and saccharification of biomass. Enzyme Microb Technol 6:503-507

17. Betancur GJV (2005) Avanços em biotecnologia de hemicelulose Para produçao de etanol por Pichia stipitis. Dissertaçao de Mestrado, Escola de Quimica da UFRJ, Rio de Janeiro.

18. Bhatia L, Paliwal S (2010) Banana peel waste as substrate for ethanol production. International J of Biotechnol and Bioeng Research 1(2):213-218

19. Bhatia L, Paliwal S (2011) Ethanol producing potential of Pachysolen tannophilus from sugarcane bagasse. International J of Biotechnol and Bioeng Research 2(2):271-276

20. Bisaria VS (1998) Bioprocessing of agro-residues to value added products. In: Martin AM (ed) Bioconversion of waste materials to industrial products, 2nd edn. Chapman and Hall, UK. pp 197-246

21. Bjerre AB, Olesen AB, Fernqvist T (1996) Pretreatment of wheat straw using combined wet oxidation and alkaline hydrolysis resulting in convertible cellulose and hemicellulose. Biotechnol Bioeng 49:568-577

22. Blotkamo PJ, Takaai M, Pemberton MS, Enert GS (1978) Enzymatic hydrolysis of cellulose and simultaneous saccharification to alcohol. AIChE Symp Ser 74:85

23. Bothast RJ, Nichols NN, Dien BS (1999) Fermentation with new recombinant organisms. Biotechnol Prog 15:867-875

24. Brooks AA (2008) Ethanol production potential of local yeast strains isolated from ripe banana peels. African J of Biotechnol 7(20):3749-3752
25. Cadoche L, Lopez GD (1989) Assessment of size reduction as a preliminary step in the production of ethanol from lignocellulosic wastes. Biol Wastes 30:153-157
26. Cao Y, Li H, Zhang Y, Zhang J, He J (2010) Structure and properties of novel regenerated cellulose films prepared from cornhusk cellulose in room temperature ionic liquids. J Appl Polymer Sci 116(1):547-554
27. Champagne P (2007) Feasibility of producing bio-ethanol from waste residues: a Canadian perspective feasibility of producing bio-ethanol from waste residues in Canada. Resour Conserv Recycl 50:211-230
28. Cheng YS, Zheng Y, Yu CW, Dooley TM, Jenkins BM, Vander Gheynst JS (2010) Evaluation of high solids alkaline pretreatment of rice straw. Appl Biochem Biotechnol 162(6):1768-1784
29. Chornet E, Vanasse C, Lemonnier JP, Overend RP (1988) Preparation and processing of medium and high consistency biomass suspensions. Research in thermochemical biomass conversion. 766-778
30. Chundawat SPS, Venkatesh B, Dale BE (2007) Effect of particle size based separation of milled corn stover on AFEX pretreatment and enzymatic digestibility. Biotechnol Bioeng 96(2):219-231
31. Coveca (2002) Comision veracruzana de comercializacion agropecuaria. Gobierno del Estado de Veracruz, Mexico.
32. Curreli N, Agelli M, Pisu B, Rescigno A, Sanjust E, Rinaldi A (2002) Complete and efficient enzymic hydrolysis of pretreated wheat straw. Process Biochem 37:937-941
33. Curreli N, Fadda MB, Rescigno A, Rinaldi AC, Soddu G, Sollai F (1997) Mild alkaline/oxidative pretreatment of wheat straw. Process Biochem 32:665-670
34. Dadi AP, Schall CA, Varanasi S (2007) Mitigation of cellulose recalcitrance to enzymatic hydrolysis by ionic liquid pretreatment. Appl Biochem Biotechnol 12:407-421
35. Das S, Schlessel DB, Ji HF, Donough JM, Wei Y (2011) Enzymatic hydrolysis of biomass with recyclable use of cellobiase enzyme

immobilized in sol–gel routed mesoporous silica. J Mol Catal B Enzym 70:49-54

36. Dashtban M, Schraft H, Qin W (2009) Fungal bioconversion of lignocellulosic residues, opportunities & perspectives. Int J Biol Sci 5(6):578-595
37. Dawson L, Boopathy R (2007) Use of post-harvest sugarcane residue for ethanol production. Biores Technol 98:1695-1699
38. De Moraes-Rocha GJ, Martin C, Soares IB, Maior AMS, Baudel HM, De Abreu CAM (2010) Dilute mixed-acid pretreatment of sugarcane bagasse for ethanol production. Biomass Bioenergy 35:663-670
39. Delmer DP, Amor Y (1995) Cellulose biosynthesis. Plant Cell 7:987-1000
40. Demain AL, Michael Newcomb WJHD (2005) Cellulase, clostridia, and ethanol. Microbiol Mol Biol Rev 69(1):124-154
41. Dhillon GS, Bansal S, Oberoi HS (2007) Cauliflower waste incorporation into cane molasses improves ethanol production using Saccharomyces cerevisiae MTCC 178. Indian J Microbiol 47:353-357
42. Dien BS, Nichols NN, O'Bryan PJ, Bothast RJ (2000) Development of new ethanologenic Escherichia coli strains for fermentation of lignocellulosic biomass. Appl Biochem Biotechnol 84–86:181-186
43. Dienes D, Egyhazi A, Reczey K (2004) Treatment of recycled fiber with trichoderma cellulases. Ind Crop Prod 20(1):11-21
44. Duenas R, Tengerdy RP, Gutierrez-Correa M (1995) Cellulase production by mixed fungi in solid-substrate fermentation of bagasse. World J of Microbiol & Botechnol 11:333-337
45. Duff SJB, Murray WD (1996) Bioconversion of forest products industry waste cellulosics to fuel ethanol: a review. Bioresour Technol 55:1-33
46. Duff SJB (1985) Cellulase and beta-glucosidase production by mixed culture of Trichoderma reesei Rut C30 and Aspergillus phoenicis. Biotechnol Lett 7:185-190
47. Ebringerova A, Hromadkova Z (2002) Effect of ultrasound on the extractibility of corn bran hemicelluloses. Ultrason Sonochem 9(4):225-229

48. ElSeoud OA, Koschella A, Fidale LC, Dorn S, Heinze T (2007) Applications of ionic liquids in carbohydrate chemistry: a window of opportunities. Biomacromolecules 8(9):2629-2647

49. Elumalia S, Thangavelu V (2010) Simultaneous saccharification and fermentation (SSF) of pretreated sugarcane bagasse using cellulose and Saccharomyces cerevisiae-kinetics and modeling. Chem Eng Res Bull 14:29-35

50. ESep (2008) A Novel Low Energy Route to Ethanol Recovery, Trans Ionics Corporation21st NREL Industry Growth Forum October 28-30, Trans Ionics Corporation. Incorporated 2000. Energy-Saving Separation.

51. Fan LT, Gharpuray MM, Lee YH (1987) Cellulose hydrolysis biotechnology monographs. Springer, Berlin. p 57

52. Fenel F, Zitting AJ, Kantelinen A (2006) Increased alkali stability in Trichoderma reesei endo-1, 4-beta-xylanase II by site directed mutagenesis. J Biotechnol 121:102-107

53. Fenel F, Leisola M, Janis J, Turunen O (2004) A de novo designed N-terminal disulphide bridge stabilizes the trichoderma reesei endo-1,4-beta-xylanase II. J Biotechnol 108:137-143

54. Ferraz A, Ana M, Cordova A (2003) Machuca: wood biodegradation and enzyme production by Ceriporiopsis subvermispora during solid-state fermentation of Eucalyptus grandis. Enzyme and Microbial Technol 32:59-65

55. Focher B, Marzett A, Crescenzi V (eds) (1991) Steam explosion techniques, fundamentals and industrial applications Gordon and Breach, Philadelphia, Pa, USA.

56. Galbe M, Zacchi G (2002) A review of the production of ethanol from softwood. Appl Microbiol Biotechnol 59:618-628

57. Gama FM, Carvalho MG, Figueiredo MM, Mota M (1997) Comparative study of cellulose fragmentation by enzymes and ultrasound. Enzyme Microb Technol 20:12-17

58. Gerchman Y, Schnitzer A, Gal R, Mirsky N, Chinkov N (2012) A simple rapid gas-chromatography flame-ionization-detector (GC-FID) method for the determination of ethanol from fermentation processes African J of. Biotechnol 11(15):3612-3616

59. Ghose TK, Bisaria VS (1979) Studies on mechanism of enzymatic hydrolysis of cellulosic substances. Biotechnol Bioeng 21:131-146

60. Ghose TK, Roychoudhury PK, Ghosh P (1984) Simultaneous saccharification and fermentation (SSF) of lignocellulosics to ethanol under vacuum cycling and step feeding. Biotechnol Bioeng 26:377-381

61. Goldstein I, Easter J (1992) an improved process for converting cellulose to ethanol. Tappi J 28:135

62. Grassick A, Murray PG, Thompson R, Collins CM, Byrnes L, Birrane G, Higgins TM, Tuohy MG (2004) Three-dimensional structure of a thermostable native cellobiohydrolase, CBH IB, and molecular characterization of the cel7 gene from the filamentous fungus, Talaromyces emersonii. Eur J Biochem 271:4495-4506

63. Gupte A, Madamwar D (1997) Solid state fermentation of lignocellulosic waste for cellulose and β-glucosidase production by co-cultivation by Aspergillus ellipticus and Aspergillus fumigatus. Biotechnol Prog 13:166-169

64. Ha MA, Apperley DC, Evans BW, Huxham IM, Jardine WG, Vietor RJ, Reis D, Vian B, Jarvis MC (1998) Fine structure in cellulose microfibrils: NMR evidence from onion and quince. Plant J 16:183-190

65. Hahn-Hagerdal B, Galbe M, Gorwa-Grauslund MF, Lidan G, Zacchi G (2006) Bio-ethanol-the fuel of tomorrow from the residues of today. Trends Biotechnol 24(12):549-556

66. Hall G, Reuter WM (2007) HPLC Analysis for the Monitoring of Fermentation Broth During Ethanol Production as a Biofuel Application brief, Perkin Elmer. http://www.perkinelmer.com/pdfs/downloads/ABR_ EthanolAsBiofuelbyHPLCAppBrief.pdf

67. Hamelinck CN, Hooijdonk GV, Faaji APC (2005) Ethanol from lignocellulosic biomass: techno-economic performance in short-, middle- and long-term. Biomass Bioenergy 28:384-410

68. Hebeish A, Ibrahim NA (2007) The impact of frontier sciences on textile industry. Colourage 54:41-55

69. Himmel ME, Ding SY, Johnson DK, Adney WS, Nimlos MR, Brady JW, Foust TD (2007) Biomass recalcitrance: engineering plants

and enzymes for biofuels production. Science 315(5813):804-807

70. Ho NWY, Chen Z, Brainard AP (1998) Genetically engineered saccharomyces yeast capable of effective cofermentation of glucose and xylose. Appl Environ Microbiol 64:1852-1859
71. Howard RL, Abotsi E, Jansen van Rensburg EL, Howard S (2003) Lignocellulose biotechnology: issues of bioconversion and enzyme production. African J of Biotechnol 2:602-619
72. Ibrahim MM, El-Zawawy WK, Abdel-Fattah YR, Soliman NA, Agblevor FA (2011) Comparison of alkaline pulping with steam explosion for glucose production from rice straw. Carbohydr Polym 83(2):720-726
73. Ikweb J, Harvey AP (2011) Intensification of bioethanol production by simultaneous saccharification and fermentation (SSF) in an oscillatory baffled reactor (OBR). Bioenergy technol. World Renewable Energy Congress, Sweden. pp 381-388
74. Imai M, Ikari K, Suzuki I (2004) High-performance hydrolysis of cellulose using mixed cellulase species and ultrasonication pretreatment. Biochem Eng J 17:79-83
75. Isitua CC, Ibeh IN (2010) Novel method of wine production from banana (Musa acuminata) and pineapple (Ananas comosus) wastes. African J of Biotechnol 9(44):7521-7524
76. Iso M, Chen B, Eguchi M, Kudo T, Shrestha S (2001) Production of biodiesel fuel from triglycerides and alcohol using immobilized lipase. J Mol Catal B: Enzym 16(1):53-58
77. Ivanova V, Hristov J, Dobreva E, AlHassan Z, Penchev I (1996) Performance of a magnetically stabilized bed reactor with immobilized yeast cells. Appl Biochem Biotechnol 59:187-198
78. Ivanova V, Petrova P, Hristov J (2011) Application in the ethanol fermentation of immobilized yeast cells in matrix of alginate/magnetic nanoparticles, on chitosan-magnetite microparticles and cellulose-coated magnetic nanoparticles. Int Rev Chem Eng 3:289-299
79. Iyer PV, Wu ZW, Kim SB, Lee YY (1996) Ammonia recycled percolation process for pretreatment of herbaceous biomass. Appl Biochem Biotechnol 57–58:121-132

80. Jacobsen SE, Wyman CE (2000) Cellulose and hemicellulose hydrolysis models for application to current and novel pretreatment processes. Appl Biochem Biotechnol 84(1–9):81-96

81. Juhasz T, Kozma K, Szengyel Z, Reczey K (2003) Production of β-glucosidase in mixed culture of Aspergillus Niger BKMF 1305 and Trichoderma reesei RUT C30. Food Technol Biotechnol 41:49-53

82. Kalogo Y, Habibi S, MacLean HL, Joshi SV (2007) Environmental implications of municipal solid waste-derived ethanol. Environ Sci Technol 41:35-41

83. Kaur J, Chandha BS, Kumar BA (2007) Purification and characterization of β-glucosidase from melanocarpus Sp. MTCC3922. Elect J Biotechnol 10:260-270

84. Keller GE, Bryan PF (2000) Process engineering moving in New directions. Chem Eng Prog 96(1):41-50

85. Kilzer FJ, Broido A (1965) Speculations on the nature of cellulose pyrolysis. Pyrodynamics 2:151-163

86. Kim KH, Hong J (2001) Supercritical CO2 pretreatment of lignocellulose enhances enzymatic cellulose hydrolysis. Bioresour Technol 77(2):139-144

87. Kim S, Dale BE (2004) Global potential bioethanol production from wasted crops and crop residues. Biomass Bioenergy 26:361-375

88. Kim TH, Lee YY (2005) Pretreatment and fractionation of corn stover by ammonia recycle percolation process. Bioresour Technol 96(18):2007-2013

89. King JW, Srinivas K (2009) Multiple unit processing using sub- and supercritical fluids. J Supercrit Fluids 47(3):598-610

90. Kovacs K, Megyeri L, Szakacsa G, Kubicekc CP, Galbeb M, Zacchi G (2008) Trichoderma atroviride mutants with enhanced production of cellulase and β-glucosidase on pretreated willow. Enzyme Microb Technol 43:48-55

91. Kuhad RC, Gupta R, Khasa YP (2010) Bioethanol production from lignocellulosic biomass: an overview. In: Lal B (ed) Wealth from waste, Teri Press, New Delhi, India.

92. Kumar P, Barrett DM, Delwiche MJ, Stroeve P (2009) Methods for pretreatment of lignocellulosic biomass for efficient hydrolysis and biofuel production. Ind Eng Chem Res 48:3713-3729

93. Kumar R, Singh S, Singh OV (2008) Bioconversion of lignocellulosic biomass: biochemical and molecular perspectives. J Ind Microbiol Biotechnol 35:377-391

94. Kuo CH, Lee CK (2009) Enhanced enzymatic hydrolysis of sugarcane bagasse by N-methylmorpholine-N-oxide pretreatment. Bioresour Technol 100(2):866-871

95. Lachenmeier DW, Godelmann R, Steiner M, Ansay B, Weigel J, Krieg G (2010) Rapid and mobile determiantion of alcoholic stregth in wine, beer, and sprits using a flow-through infrared sensor. Chem Cent J 4:5-15

96. Lavarack BP, Giffin GJ, Rodman D (2002) The acid hydrolysis of sugarcane bagasse hemicellulose to produce xylose, arabinose, glucose, and other products. Biomass Bioenerg 23:367-380

97. Lelkes Z, Szitkai Z, Rev E, Fonyo Z (2000) Rigorous MINLP model for ethanol dehydration system. Computers and Chem Eng 24:1331-1336

98. Li Q, He YC, Xian M (2009) Improving enzymatic hydrolysis of wheat straw using ionic liquid 1-ethyl-3-methyl imidazolium diethyl phosphate pretreatment. Bioresour Technol 100(14):3570-3575

99. Li YL, Li DC, Teng FC (2006) Purification and characterization of a cellobiohydrolase from the thermophilic fungus Chaetomium thermophilus CT2. Wei Sheng Wu Xue Bao 46:143-146

100. Lin Y, Tanaka S (2006) Ethanol fermentation from biomass resources: current state and prospects. Appl Microbiol Biotechnol 69:627-642

101. Lu GQ, Huang HH, Zhang DP (2006) Application of near-infrared spectroscopy to predict sweet potato starch thermal properties and noodle quality. J. Zhejiang University (Eng. Sci.) 6:475-481

102. Lynd LR, van Zyl WH, McBride JE, Laser M (2005) Consolidated bioprocessing of cellulosic biomass: an update. Curr Opin Biotechnol 16:577-583

103. Lynd LR, Jin H, Michels JG, Wyman CE, Dale B (2003) Bioenergy: background, potential, and policy. Center for Strategic & International Studies, Washington, D.C..

104. Lynd LR, Weimer PJ, van Zyl WH, Pretorius IS (2002) Microbial cellulose utilization: fundamentals and biotechnology. Microbiol Mol Biol Rev 66:506-577

105. MacDonald DG, Bakhshi N, Mathews JF, Roychowdhury A, Bajpai P, Moo-Young M (1983) Alkaline treatment of corn stover to improve sugar production by enzymatic hydrolysis. Biotechnol Bioeng 25(8):2067-2076

106. McIntosh S, Vancov T (2010) Enhanced enzyme saccharification of sorghum bicolor straw using dilute alkali pretreatment. Bioresour Technol 101(17):6718-6727

107. Maheshwari DK, Gohade S, Paul J, Verma A (1994) A paper mill sludge as a potential source for cellulase production by Trichoderma reesei QM9123 and Aspergillus Niger using mixed cultivation. Carbohydr Polym 23:161-163

108. Maheshwari R, Bharadwaj G, Bhat MK (2000) Thermophilic fungi: their physiology and enzymes. Microbiol Mol Biol Rev 64:461-488

109. Mandels M, Weber J, Parizek R (1971) Enhanced cellulase production by a mutant of Trichoderma viride. Appl Microbiol 21:152-154

110. Martin C, Klinke HB, Thomsen AB (2007) Wet oxidation as a pretreatment method for enhancing the enzymatic convertibility of sugarcane bagasse. Enzyme and Microbial Technol 40(3):426-432

111. Martin C, Thomsen AB (2007) Wet oxidation pretreatment of lignocellulosic residues of sugarcane, rice, cassava and peanuts for ethanol production. J of Chemical Technol and Biotechnol 82(2):174-181

112. Martin C, Marcet M, Thomsen AB (2008) Comparison between wet oxidation and steam explosion as pretreatment methods for enzymatic hydrolysis of sugarcane bagasse. BioResources 3(3):670-683

113. Martinez AT, Speranza M, Ruiz-Duenas FJ, Ferreira P, Camarero S, Guillen F, Martinez MJ, Gutierrez A, del Rio JC (2005) Biodegradation of lignocellulosics: microbial, chemical, and enzymatic aspects of the fungal attack of lignin. Int Microbiol 8:195-204

114. McMillan JD (1994) Pretreatment of lignocellulosic biomass. In: Himmel ME, Baker JO, Overend RP (eds) Enzymatic conversion of biomass for fuels production, American Chemical Society, Washington, DC. pp 292-324

115. Meena K, Raja TK (2004) Immobilization of yeast invertase by gel entrapment. Ind J of Biotechnol 3:606-608

116. Millet MA, Baker AJ, Scatter LD (1976) Physical and chemical pretreatment for enhancing cellulose saccharification. Biotech Bioeng Symp 6:125-153

117. Montane D, Farriol X, Salvado J, Jollez P, Chornet E (1998) Application of stream explosion to the fractionation and rapid vapour-phase alkaline pulping of wheat straw. Biomass Bioenergy 14:261-276

118. Moredo N, Lorenzo M, Domínguez A, Moldes D, Cameselle C, Sanroman A (2003) Enhanced ligninolytic enzyme production and degrading capability of Phanerochaete chrysosporium and Trametes versicolor. World J Microb Biotechnol 19:665-669

119. Morohoshi N (1991) Chemical characterization of wood and its components. In: Hon DNS, Shiraishi N (eds) Wood and cellulosic chemistry, Marcel Dekker, Inc, New York, USA. pp 331-392

120. Mosier N, Wyman CE, Dale BD, Elander RT, Lee YY, Holtzapple M, Ladisch CM (2005) Features of promising technologies for pretreatment of lignocellulosic biomass. Bioresour Technol 96:673-686

121. Murray P, Aro N, Collins C, Grassick A, Penttila M, Saloheimo M, Tuohy M (2004) Expression in Trichoderma reesei and characterisation of a thermostable family 3 beta-glucosidase from the moderately thermophilic fungus Talaromyces emersonii. Protein Expr Purif 38:248-257

122. Ohta K, Beall DS, Meija JP, Shanmugam KT, Ingram LO (1991) Metabolic engineering of Klebsiella oxytoca M5A1 for ethanol production from xylose and glucose. Appl Environ Microbiol 57:2810-2815

123. Okada H, Mori K, Tada K, Nogawa M, Morikawa Y (2000) Identification of active site carboxylic residues in Trichoderma reesei endoglucanase Cel12A by site-directed mutagenesis. J Mol Catal 10:249-255

124. Palacios-Orueta A, Chuvieco E, Parra A, Carmona-Moreno C (2005) Biomass burning emissions: a review of models using remote-sensing data. Environ Monit Assess 104:189-209

125. Panda T, Bisaria VS, Ghose TK (1983) Studies on mixed fungal culture for cellulase and hemicellulase production. Part 1. Optimization of medium for the mixed culture of Trichoderma reesei DI-6 and Aspergillus wentii Pt 2804. Biotechnol Lett 5:767-772

126. Pappas C, Tarantilis PA, Daliani I, Mavromoustakos T, Polissiou M (2002) Comparison of classical and ultrasound-assisted isolation procedures of cellulose from kenaf (Hibiscus cannabinus L.) and eucalyptus (Eucalyptus rodustrus Sm.). Ultrason Sonochem 9(1):19-23

127. Pere J, Puolakka A, Nousiainen P, Buchert J (2001) Action of purified Trichoderma reesei cellulases on cotton fibers and yarn. J Biotechnol 89(2–3):247-255

128. Perepelkin KE (2007) Lyocell fibres based on direct dissolution of cellulose in N-methylmorpholine N-oxide: development and prospects. Fibre Chemistry 39(2):163-172

129. Perez VH, Reyes AF, Justo OR, Alvarez DC, Alegre RM (2007) Bioreactor coupled with electromagnetic field generator: effects of extremely Low frequency electromagnetic fields on ethanol production by Saccharomyces cerevisiae. Biotechnol Progress 23:1091-1094

130. Persson T, Matusiak M, Zacchi G, Jonsson A-S (2006) Extraction of hemicelluloses from process water from the production of masonite. Desalination 199:411-412

131. Philippidis GP, Smith TK, Wyman CE (1993) Study of the enzymatic hydrolysis of cellulose for production of fuel ethanol by the simultaneous saccharification and fermentation process. Biotechnol Bioeng 41:846-853

132. Qin Y, Wei X, Song X, Qu Y (2008) Engineering endoglucanase II from Trichoderma reesei to improve the catalytic efficiency at a higher pH optimum. J Biotechnol 135:190-195

133. Ragauskas AJ, Williams CK, Davison BH, Britovsek G, Cairney J, Eckert CA, Frederick WJ, Hallett JP, Leak DJ, Liotta CL, Mielenz JR, Murphy R, Templer R, Tschaplinski T (2006) The path forward for biofuels and biomaterials. Science 311:484-489

134. Rahman Z, Shida Y, Furukawa T, Suzuki Y, Okada H, Ogasawara W, Morikawa Y (2009) Application of Trichoderma reesei cellulase and xylanase promoters through homologous recombination for enhanced production of extracellular beta-glucosidase I. Biosci Biotechnol Biochem 73:1083-1089

135. Raposo S, Pardao JM, Diaz I, Costa MEL (2009) Kinetic modelling of bioethanol production using agro-industrial by-products. Int J of Energy Env 3(1):8

136. Rolz C (1986) Ultrasound effect of enzymatic saccharification. Biotech Letters 8(2):131-136

137. Rosenau T, Potthast A, Sixta H, Kosma P (2001) The chemistry of side reactions and byproduct formation in the system NMMO/cellulose (lyocell process). Prog Polym Sci 26(9):1763-1837

138. Rosgaard L, Pedersen S, Cherry JR, Harris P, Meyer AS (2006) Efficiency of new fungal cellulase systems in boosting enzymatic degradation of barley straw lignocellulose. Biotechnol Prog 22:493-498

139. Ruffell J, Levie B, Helle S, Duff S (2010) Pretreatment and enzymatic hydrolysis of recovered fibre for ethanol production. Bioresour Technol 101(7):2267-2272

140. Saha BC (2003) Hemicellulose bioconversion. J Ind Microbiol Biotechnol 30:279-291

141. Saha BC, Cotta MA (2007) Enzymatic saccharification and fermentation of alkaline peroxide pretreated rice hulls to ethanol. Enzyme Microbiol Technol 41:528-532

142. Saha BC, Cotta MA (2011) Continuous ethanol production from wheat straw hydrolysate by recombinant ethanologenic Escherichia coli strain FBR5. Appl Microbiol Biotechnol 90:477-487

143. Sakai Y, Tamiya Y, Takahashi F (1994) Enhancement of ethanol formation by immobilized yeast containing iron powder or Ba-ferrite due to eddy current or hysteresis. J Ferment Bioeng 77:169-172

144. Sang-Mok L, Koo YM (2001) Pilot-scale production of cellulase using Rut C-30 in fed-batch mode. J Microbiol Biotechnol 11(2):229-233

145. Sanchez C (2009) Lignocellulosic residues: biodegradation and bioconversion by fungi. Biotechnol Adv 27(2):185-194

146. Sasikumar E, Viruthagiri T (2010) Simultaneous saccharification and fermentation (SSF) of sugarcane bagasse - kinetics and modeling. Int J of Chemical and Biological Eng 3(2):57-64

147. Schmidt AS, Mallon S, Thomsen AB, Hvilsted S, Lawther JM (2002) Comparison of the chemical properties of wheat straw and beech fibers following alkaline wet oxidation and laccase treatments. J of Wood Chem and Technol 22(1):39-53

148. Schurz J (1978)Ghose TK (ed) Bioconversion of cellulosic substances into energy chemicals and microbial protein symposium proceedings, IIT, New Delhi. p 37

149. Scott WE, Gerber P (1995) Using ultrasound to deink xerographic waste. Tappi J 78:125-130

150. Senthilkumar V, Gunasekaran P (2005) Bioethanol production from cellulosic substrate: engineered bacteria and process integration challenges. J of Sci and Inds Resrch 64:845-853

151. Shallom D, Shoham Y (2003) Microbial hemicellulases. Curr Opin Microbiol 6:219-228

152. Sharma N, Kalra KL, Oberoi HS, Bansal S (2007) Optimization of fermentation parameters for production of ethanol from kinnow waste and banana peels by simultaneous saccharification and fermentation. Ind J Microbiol 47:310-316

153. Sharma K, Sharma SP, Lahiri SC (2009) Novel method for identification and quantitation of methanol and ethanol in alcoholic beverages by gas chromatography-Fourier transform infrared spectroscopy and horizontal attenuated total reflectance-Fourier transform infrared spectroscopy. J AOAC Int 92:518-526

154. Shafiei M, Karimi K, Taherzadeh MJ (2010) Pretreatment of spruce and oak by N-methylmorpholine-N-oxide (NMMO) for efficient conversion of their cellulose to ethanol. Bioresour Technol 101(13):4914-4918

155. Shafizadeh F, Bradbury AGW (1979) Thermal degradation of cellulose in air and nitrogen at low temperatures. J Appl Poly Sci 23:1431-1442

156. Shaw AJ, Podkaminer KK, Desai SG, Bardsley JS, Rogers SR, Thorne PG, Hogsett DA, Lynd LR (2008) Metabolic engineering of a thermophilic bacterium to produce ethanol at high yield. Proc Natl Acad Sci U S A 105(37):13769-13774

157. Shi J, Chinn MS, Sharma-Shivappa RR (2008) Microbial pretreatment of cotton stalks by solid state cultivation of phanerochaete chrysosporium. Bioresour Technol 99:6556-6564

158. Sills DL, Gossett JM (2011) Assessment of commercial hemicellulases for saccharification of alkaline pretreated perennial biomass. Bioresour Technol 102(2):1389-1398

159. Sims R (2003) Biomass and resources bioenergy options for a cleaner environment in developed and developing countries. Elsevier Science London, UK.

160. Sipos B, Reczey J, Somorai Z, Kadar Z, Dienes D, Reczey K (2009) Sweet sorghum as feedstock for ethanol production: enzymatic hydrolysis of steam pretreated bagasse. Appl Biochem Biotechnol 153:151-162

161. Sipos B, Kreuger E, Svensson S-E, Reczey K, Bjornsson L, Zacchi G (2010) Steam pretreatment of dry and ensiled industrial hemp for ethanol production. Biomass Bioenergy.

162. Sohel MI, Jack MW (2010) Therrmodynamic analysis of lignocellulosic biofuel production via a biochemical process: guiding technology selection and research focus. Bioresour Technol.

163. Soloveva IV, Okunev ON, Velkov VV, Koshelev AV, Bubnova TV, Kondrateva EG, Skomarovskii AA, Sinitsyn AP (2005) The selection and properties of Penicillium verruculosum mutants with enhanced production of cellulases and xylanases. Mikrobiologiia 74:172-178

164. Soto ML, Dominguez H, Nunez MJ, Lema JM (1994) Enzymatic saccharification of alkali-treated sunflower hulls. Bioresour Technol 49(1):53-59

165. Stenberg K, Bollok M, Reczey K, Galbe M, Zacchi G (2000) Effect of substrate and cellulase concentration on simultaneous saccharification and fermentation of steam-pretreated softwood for ethanol production. Biotechnol Bioeng 68:204-210
166. Sternberg D (1976) Production of cellulase by trichoderma. Biotechnol Bioeng Symp 6:35-53
167. Stockton BC, Mitchell DJ, Grohmann K, Himmel ME (1991) Optimum β-D-glucosidase supplementation of cellulase for efficient conversion of cellulose to glucose. Biotechnol Lett 13:57-62
168. Sukumaran RK, Singhania RR, Pandey A (2005) Microbial cellulases—production, applications and challenges. Jr of Sci and Inds Resrch 64(11):832-844
169. Sun FB, Cheng HZ (2007) Evaluation of enzymatic hydrolysis of wheat straw pretreated by atmospheric glycerol autocatalysis. J Chem Technol Biotechnol 82:1039-1044
170. Sun RC, Tomkinson J (2002) Comparative study of lignins isolated by alkali and ultrasound-assisted alkali extractions from wheat straw. Ultrason Sonochem 9(2):85-93
171. Sun YE, Cheng J (2002) Hydrolysis of lignocellulosic materials for ethanol production: a review. Bioresour Technol 83(1):1-11
172. Swatloski RP, Spear SK, Holbrey JD, Rogers RD (2002) Dissolution of cellulose with ionic liquids. J Am Chem Soc 124:4974-4975
173. Szczodrak J (1989) The use of cellulases from a j-glucosidasc hyperproducing mutant of trichoderma reesei in simultaneous saccharification and fermentation of wheat straw. Biorechnol Bioeng 9:1112
174. Szijarto N, Kadar Z, Varga E, Thomsen AB, Costa-Ferreira M, Reczey K (2009) Pretreatment of reed by wet oxidation and subsequent utilization of the pretreated fibers for ethanol production. Appl Biochem Biotechnol 155(1–3):386-396
175. Szitkai Z, Lelkes Z, Rev E, Fonyo Z (2002) Optimization of hybrid ethanol dehydration systems. Chem Eng and Processing 41:631-646
176. Taherzadeh MJ, Karimi K (2008) Pretreatment of lignocellulosic wastes to improve ethanol and biogas production: a review. Int J Mol Sci 9:1621-1651

177. Takagi M, Abe S, Suzuki S, Emert GH, Yata N (1977) A method for production of alcohol directly from cellulose using cellulose and yeast. Proceedings, Bioconversion Symposium, IIT, Delhi. pp 551-571

178. Talebnia F, Karakashev D, Angelidaki I (2010) Production of bioethanol from wheat straw: an overview on pretreatment, hydrolysis and fermentation. Bioresour Technol 101(13):4744-4753

179. Teymouri F, Laureano-Perez L, Alizadeh H, Dale BE (2004) Ammonia fiber explosion treatment of corn stover. Appl Biochem and Biotechnol, Part A 115(1–3):951-963

180. Tsuyomoto M, Teramoto A, Meares P (1997) Dehydration of ethanol on a pilot plant scale, using a new type of hollow-fiber membrane. J of Membrane Sci 133:83-94

181. Turner P, Mamo G, Karlsson EN (2007) Potential and utilization of thermophiles and thermostable enzymes in biorefining. Microb Cell Fact 6:9

182. Viikari L, Alapuranen M, Puranen T, Vehmaanpera J, Siika-Aho M (2007) Thermostable enzymes in lignocellulose hydrolysis. Adv Biochem Eng Biotechnol 108:121-145

183. Vinatoru M, Toma M, Mason TJ (1999) Ultrasonically assisted extraction of bioactive principles from plants and their constituents. In: Mason TJ (ed) Advances in sonochemistry, 5th edn. JAI Press, London. pp 209-248

184. Wang ML, Choong YM, Su NW, Lee MH (2003) A rapid method for determination of ethanol in alcholic beverages using capillary Gas chromatography. Jr of Food and Drug Analysis 11(2):1330-140

185. Wang ZM, Li L, Xiao KJ, Wu JY (2009) Homogeneous sulfation of bagasse cellulose in an ionic liquid and anticoagulation activity. Bioresour Technol 100(4):1687-1690

186. Wati L, Kumari S, Kundu BS (2007) Paddy straw as a substrate for ethanol production. Ind Jr Of Microbiol 47:26-29

187. Wen Z, Liao W, Chen S (2004) Hydrolysis of animal manure lignocellulosics for reducing sugar production. Bioresour Technol 91:31-39

188. Wingren A, Galbe M, Zachhi G (2003) Techno-economic evaluation of producing ethanol from softwood: comparision of SSF and SHF and identification of bottle necks. Biotechnol Prog 19:1086-1093

189. Wood BE, Aldrich HC, Ingram LO (1997) Ultrasound stimulates ethanol production during the simultaneous saccharification and fermentation of mixed waste office paper. Biotechnol Prog 13(3):232-237

190. Wright JD (1998) Ethanol from biomass by enzymatic hydrolysis. Chem Eng Prog 84(8):62-74

191. Wyman CE, Spindler DD, Grohmann K, Lastick SM (1986) Simultaneous saccharification and fermentation of cellulose with the yeast Brettanomyces clausenii. Biotechnol and Bioeng Symp No 17:221-238

192. Wyman CE, Hinman ND (1990) Ethanol. Fundamentals of production from renewable feedstocks and use as a transportation fuel. Appl Biochem Biotechnol 24:735-753

193. Yamashita Y, Sasaki C, Nakamura Y (2010) Effective enzyme saccharification and ethanol production from Japanese cedar using various pretreatment methods. J Biosci Bioeng 110:79-86

194. Yang B, Wyman CE (2008) Pretreatment: The key to unlocking low cost cellulosic ethanol. Biofuel Bioprod Biorefin 2:26-40

195. Yu G, Yano S, Inoue H, Inoue S, Endo T, Sawayama S (2010) Pretreatment of rice straw by a hot-compressed water process for enzymatic hydrolysis. Appl Biochem Biotechnol 160(2):539-551

196. Zaldivar GJ, Nielsen J, Olsson L (2001) Fuel ethanol production from lignocellulose: a challenge for metabolic engineering and process integration. Appl Microbiol Biotechnol 56:17-34

197. Zhang L, Zhao H, Gan M, Jin Y, Gao X, Chen Q, Guan J, Wang Z (2011) Application of simultaneous saccharification and fermentation (SSF) from viscosity reducing of raw sweet potato for bioethanol production at laboratory, pilot and industrial scales. Bioresour Technol 102:4573-4579

198. Zhang M, Eddy C, Deanda K, Finkelstein M, Picataggio S (1995) Metabolic engineering of a pentose metabolism pathway in ethanologenic Zymomonas mobilis. Science 267:240-243

199. Zhang Y-HP, Lynd LR (2004) Toward an aggregated understanding of enzymatic hydrolysis of cellulose: non-complexed cellulase systems. Biotechnol Bioeng 88:797-824

200. Zhao H, Jones CL, Baker GA, Xia S, Olubajo O, Person VN (2009) Regenerating cellulose from ionic liquids for an accelerated enzymatic hydrolysis. J Biotechnol 139(1):47-54

201. Zhao Y, Wang Y, Zhu JY, Ragauskas A, Deng Y (2008) Enhanced enzymatic hydrolysis of spruce by alkaline pretreatment at low temperature. Biotechnol and Bioengi 99(6):1320-1328

202. Zhu JY, Pan XJ, Wang GS, Gleisner R (2009) Sulfite pretreatment for robust enzymatic saccharification of spruce and red pine. Bioresour Technol 100:2411-2418

203. Zhu S, Wu Y, Chen Q, Yu Z, Wang C, Jin S, Ding Y, Wu G (2006) Dissolution of cellulose with ionic liquids and its application: a mini-review. Green Chem 8:325-327

204. Zhu S, Wu Y, Yu Z, Zhang X, Wang C, Yu F, Jin S (2006) Production of ethanol from microwave-assisted alkali pretreated wheat straw. Process Biochem 41:869-873

Chapter 4

Effects of Wax Treatment on Quality and Postharvest Physiology of Pineapple Fruit in Cold Storage

Huigang Hu[1,2], Xueping Li[1], Chen Dong[3], and Weixin Chen[1]

[1]College of Horticulture, South China Agricultural University, Guangzhou 510642, China

[2]South Subtropical Crop Research Institute, Chinese Academy of Tropical Agricultural Science, Zhanjiang 524091, China

[3]Department of Life Science and Technology, Yangjiang Vocational and Technical College, Yangjiang 529566, China

ABSTRACT

Pineapple (Ananas comosus (L.) Merr.) is an important fruit crop grown in tropical and subtropical areas. Cold storage is one of methods for prolonging postharvest storage life for pineapple fruit. A major problem of this method is that low temperature causes chilling injury symptoms in fruits and deterioration of their quality and nutritional values. This study aimed to find a new way to effectively alleviate chilling injury and maintain fruit quality during cold storage. Thus, the effects of two types of waxing treatment (Sta-Fresh 2952 wax and Sta-Fresh 7055 wax) of pineapple fruits of cultivar'Paris' (a major cultivar for pineapple production) on alleviating chilling injury and their physiological responses during cold storage were examined. Sta-Fresh 2952 wax (60 g/l) was more effective in: (1) alleviating chilling injury, which delayed the changes in firmness, flesh color, weight loss and soluble protein content; (2) decreasing titratable acidity, total soluble solids, cell membrane permeability and malondialdehyde content when compared with those in control. This waxing also improved total sugars and the contents of ascorbic acid in pineapple fruits. These results suggested that this treatment might be a useful technique to alleviate chilling injury and maintain fruit quality during cold storage.

INTRODUCTION

Pineapple (Ananas comosus (L.) Merr.) is an important fruit crop grown in many tropical and subtropical countries. Fresh pineapple fruit is perishable (Chen and Paull, 2001; Avallone et al., 2003; Soares et al., 2005; Wilsonwijeratnam et al., 2005; Ko et al., 2006) and cold storage is the main method to slow the product deterioration in terms of consumer perception and nutritional value (Zhang et al., 2009; Cantín et al., 2010). However, low temperature results in chilling injury symptoms in pineapple fruits during or after cold storage (Selvarajah et al., 2001; Zhou et al., 2003). Chilling injury hygrophanous flesh in pineapple fruits and is the major is manifested as internal browning (black heart) and postharvest limitation for the pineapple industry, as fruits rapidly develop blackheart following low temperature, which severely restricts refrigerated seafreight export (Selvarajah et al., 2001). Thus, postharvest treatments that can alleviate chilling injury and extend shelf-life of pineapple fruits are urgently needed. The development of effective methods to alleviate chilling injury in pineapple fruits has been widely reported and these include heat treatment and controlled atmosphere (Wilsonwijeratnam et al., 2005; Nimitkeatkai et al., 2006).

However, these methods do not eliminate significantly chilling injury. At present, wax (edible coatings) have been used as an effective technology to increase the quality of postharvest fruits and vegetables (Qiuping and Wenshui, 2007; Fan et al., 2009; Tzoumaki et al., 2009; Tietel et al., 2010). Coatings could effectively retard the loss of the water, titratable acidity and ascorbic acid of sweet cherries (Dang et al., 2010). Waxing could improve firmness, titratable acidity, ascorbic acidity andthe water content for Murcott tangor stored at 15°C for 56 days (Chien et al., 2007). Waxing, acting as semipermeable barriers, may be an effective method to alleviate chilling injury (Meng et al., 2008; Ahmed et al., 2009). However, only two studies reported the application of waxing of pineapple fruits during cold storage, with emphasis on chilling injury symptoms, but less attention to coating-induced quality alterations and their physiological responses in the fruits (Paul and Rohrbach, 1985; Wijeratnam et al., 2006).

Meanwhile, the main pineapple cultivar grown in the world is 'Smooth Cayenne', and in China, the most popular cultivar is 'Comte de Paris', accounting for more than 80% of the total pineapple production (Li et al., 2011). However, few data exist extending storage life of the cultivar 'Comte de Paris'. The objective of this study was to find a new method for extending storage life of pineapple cultivar 'Comte de Paris' and to investigate the method of how to alleviate chilling injury and maintain fruit quality during cold storage. Thus, this study was performed to comprehensively investigate the effects of application of a novel wax formulation on the changes in several parameters related to quality and life extend of pineapple fruits (weight loss, firmness, color saturation, total soluble solids (TSS), soluble sugars, titratable acidity (TA), ascorbic acid (AsA) and soluble proteins) and on their physiological activities including the changes in relative leakage rate, malondialdehyde (MDA) and membrane integrity in response to cold storage of the chilling sensitive pineapple fruits. The results obtained from this study indicated that Sta-Fresh 2952 wax treatment is more effective in alleviating chilling injury and in maintaining the quality of these pineapple fruits.

MATERIALS AND METHODS

Plant material and experimental design Pineapple fruits (A. comosus cv. Comte de Paris) (at about 3 matured stages) were selected on the basis of the uniformity of the color and size from a commercial grower in Zhanjiang, China. The stage of ripeness was determined by visual assessment of the shell (Selvarajah et al., 2001). The scale ranges from 0 to 5: 0, all eyes are totally green; 1, < 20% of the eyes are predominantly yellow; 2, 20 to 40% of the eyes are tinged with yellow; 3, up to 65% of the eyes are predominantly yellow; 4, 65 to 90% of the eyes are fully yellow; 5, > 90% of the eyes are fully yellow and no more than 20% of the eyes are reddish orange. All fruits were cleaned and soaked in 0.05% (w/v) Iprodione solution (Kuaida, Jiangsu, China) for 2 min to eliminate potential microbes. Afterwards, the treated fruits were divided into seven groups; each group (10 fruits) was placed in a clean plastic box. The fruits of one group were dipped in water (as control). The fruits of six groups were respectively treated with StaFresh 2952 (FMC) wax solution at 30, 60 and 90 g/l and Sta-Fresh 7055 (FMC) wax solutions at 40, 80 and 120 g/l. After being air dried, the samples were placed in polyethylene bags (0.04 mm), stored at 90% relative humidity (RH) and 7°C for 21 day and transferred to 25°C for 3 day to simulate shelf conditions for chilling injury and quality evaluation. Five fruits from each box were randomly sampled to determine quality characteristics of fruits after storage periods and then the optimal treatment condition was chosen.

Forty (40) fruits were treated with the optimal treatment and 40 fruits were dipped in water (as control). All fruits were stored at 7°C and some samples were taken at intervals. After 21 day storage at 7°C, the residual samples were transferred to 25°C for three days storage. Five fruits from each box were randomly sampled every 7 days in cold storage and each day in room temperature (25°C) storage to determine physical and biochemical changes during storage periods

Chilling Injury Index (CI) Determination

The fruits were cut longitudinally in half and the CI was determined. For each fruit, CI intensity was scored from 0 to 5 according to the percentage of flesh affected (0, free from CI; 1 to 5: 10, 10 to 25, 25 to 50, 50 to 75 and >75% of the flesh discolored, respectively) (Selvarajah et al., 2001). The average CI was calculated for each group of fruit.

Firmness

Fruit firmness was measured at six equational regions of the flesh of each fruit at different storage times, using a penetrometer (Instron5542) fitted with a 8 mm diameterflat probe and results were expressed as N.

Color Assessment

Readings with colorimeter were randomly taken at six different locations on each pineapple fruit, using a total of five fruits from each group. Color factors Chroma (C) and Hue angle (H) values of each fruit were directly read with a CR-300 colorimeter (Minolta, Ramsey Corp. NY) (Nunes et al., 1995).

Soluble Sugar and Titratable Acidity

Soluble sugar was measured by phenol-sulfuric acid method described previously (Masuko et al., 2005). Titratable acidity was determined by titrating fruit juice to pH 8.1 with 0.1 mol l^{-1} NaOH. The results were expressed as mmol of hydrogenion concentration per 100 g of fresh weight (Jin et al., 2009).

Total Soluble Solid

Total soluble solid in the extracted juice of fruits was measured by a refractometer (ATAGO (Brix = 0 to 32%)) and the results were expressed as %Brix (Cheour et al., 1991).

Ascorbic Acid

The ascorbic acid of pineapple fruits was measured by 2, 6-dichlorophenolindophenol titration according to the China National Analysis Standard (GB/T 6195-1986). Briefly, flesh (10 g) from each fruit was immediately homogenized in 50 ml of 0.02 g/ml oxalic acid solution and then centrifuged at 15000 × g at 4°C for 15 min. 10 ml supernatant was titrated to a permanent pink color with 0.1% 2, 6- dichlorophenolindophenol titration. Ascorbic acid concentration was calculated according to the titration volume of 2, 6-dichlorophenolindophenol and expressed as milligram per ml.

Weight Loss

Fruit weight was measured just after harvest and at the different sampling dates. The result was expressed as percentage of weight loss relative to the initial value (taken as 100%) (Sato-Zamora et al., 2005).

Table 1: Effects of different wax treatments on chilling injury index, firmness, color saturation, contents of soluble sugars and titratable acidity of pineapple fruits after 21 day storage at 7°C followed 3 day storage at 25°C

Treatment type (g/l)	Chilling index	Firmness (N)	Color saturation (Chroma)	Soluble sugar (μg.g-1)	Titratable acidity (mmol.100 g-1)
Control	3.8a	14.61c	24.89b	94.45b	13.19a
Sta-Fresh 2952 (30)	2.3b	18.22a	28.07a	95.73ab	11.92b
Sta-Fresh 2952 (60)	1.7c	17.39ab	29.20a	98.07a	10.04c
Sta-Fresh 2952 (90)	2.1bc	15.55bc	28.02a	97.10ab	10.12c
Sta-Fresh 7055 (40)	2.6b	16.45bc	26.56ab	100.44a	12.53ab
Sta-Fresh 7055 (80)	2.5b	18.30a	26.22ab	81.12c	10.72c
Sta-Fresh 7055 (120)	2.7b	18.33a	28.67a	91.57b	10.34c

Means in the same column followed by different letters differs ($P < 0.05$) according to the least significant difference multiple range test.

Relative Leakage Rate and Malondialdehyde (MDA)

Membrane permeability, expressed by relative leakage rate was determined according to the method of Zheng and Tian (2006). MDA content was determined by the thiobarbituric acid (TBA) reaction (Dipierro and Leonardis, 1997).

Protein Content

Protein content in the enzyme extracts was determined according to the Bradford (1976) method, using bovine serum albumin as a standard.

Statistical Analysis

All data were at least repeated once and analyzed statistically by ANOVA and mean differences estimated by Duncan's new multi

RESULTS

Effect of Different Wax Treatments on Chilling Injury Index, Firmness, Color Saturation, Contents of Soluble Sugar and Titratable Acidity of Pineapple Fruits

After being stored at 7°C for 21 day and transferred to 25°C for 3 day, the pineapple fruits treated with wax exhibited significantly ($P < 0.05$) lower levels of chilling injury index (CI) than that of the control (Table 1), suggesting that wax may prevent chilling injury. The index in fruits of the 2th treatment was the lowest in all samples (Table 1). Titratable acidity (TA) levels in wax-treatments were lower than those in control (Table 1). There were significant differences in the levels of soluble sugar between wax-treatments and control (Table 1). Flesh firmness in wax-treatments was higher than that in control (Table 1), indicating that wax slowed down fruit softening process as well as pectin degradation. The color saturation for the wax-treatments was significantly higher than that in control ($P < 0.05$) and the highest color saturation of fruits was in the 2th treated fruits (Table 1). Therefore, a lower chilling index in the 2th treated fruits was observed. The 2th treatment was the optimal treatment.

Chilling Injury Index

ymptoms of chilling injury (CI), including internal browning and flesh mealiness were assessed visually. As shown in Figure 1, the CI of the fruits in control was higher than that in wax-treatment and the CI of the waxtreatment was significantly lower than that in control ($P < 0.05$) after the 14th day of storage. The results indicated that wax treatment could significantly decrease internal browning and flesh mealiness symptoms of pineapple fruits.

Firmness

Loss of firmness is one of the main factors limiting quality and the postharvest shelf-life of fruits and vegetables. Figure 2 shows that the firmness in control fruits decreased gradually along with increased storage time. Throughout storage, the loss of firmness of control was significantly greater than that of wax-treatment (Figure 2), indicating that wax delayed the decline of firmness.

Color

The color change of pineapple flesh can serve as an indicator of chilling injury and fruit quality. As shown in Figure 3a, during color development, the chroma increased slightly from harvest to the 7th day of storage and then decreased sharply up to the 22th day. It is clear that wax significantly reduced the decrease of the chroma during the storage period compared to that of the control ($P < 0.05$). Chroma is the intensity or purity of the hue and it shows the brightness of fruits. Therefore, wax treatment diminished the color brightness in comparison with control.

Figure 3b shows that after 3 day of storage, hue angle was higher in wax-treatment than in control and higher hue angle showed less browning and better quality based on Nunes et al. (1995), who reported less browning in fruits with higher hue angle. The effect of wax may be due to decreased respiration rate, which prevents fruit senescence during storage.

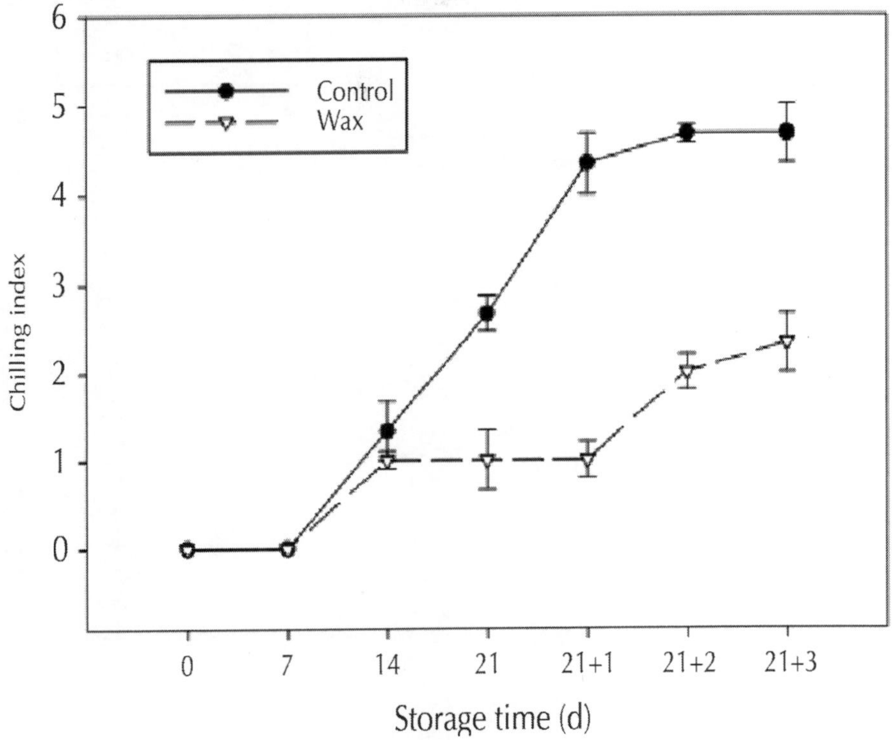

Figure 1: Effects of Sta-Fresh 2952 wax (60 g/l concentration) treatment on chilling index of pineapple fruits during storage at 7°C for 21 day and then transferred to 25°C for 3 day. "0-21" represent 21 day storage at 7°C and "21+ n" (n= 1, 2, 3) mean 21 day storage at 7°C followed by n day storage at 25°C. Values are the mean of five replicates and error bars represent the standard deviation.

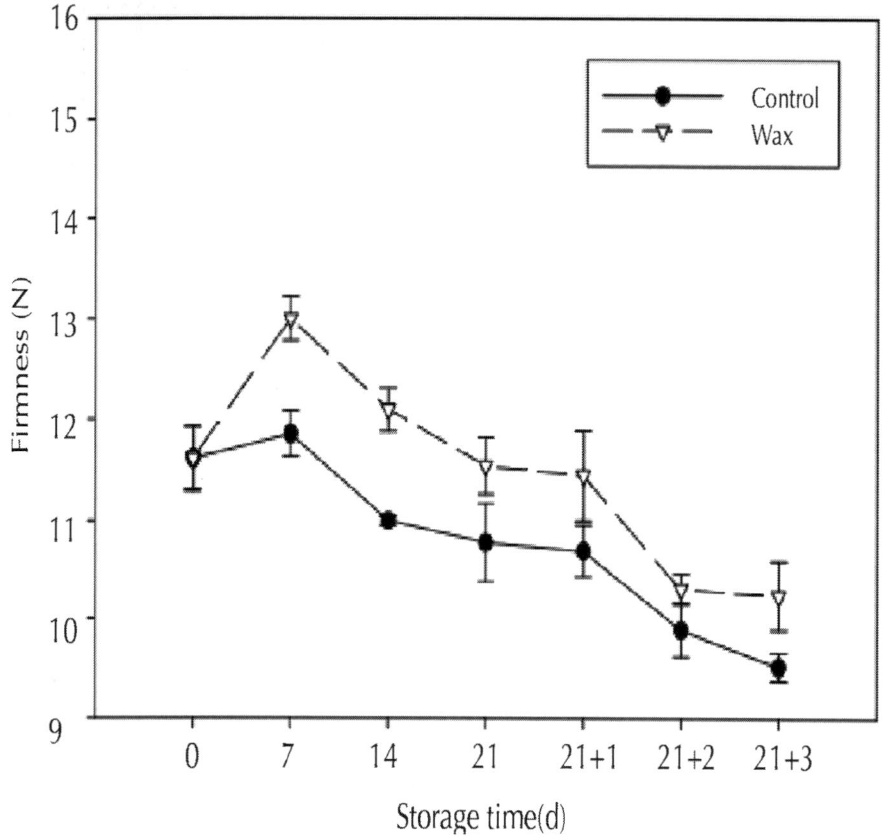

Figure 2: Effects of Sta-Fresh 2952 wax (60 g/l concentration) treatment on firmness of pineapple fruits during storage at 7°C for 21 day and then transferred to 25°C for 3 day. "0-21" represent 21 day storage at 7°C and "21+ n" (n= 1, 2, 3) mean 21 day storage at 7°C followed by n day storage at 25°C. Values are the mean of five replicates and error bars represent the standard deviation.

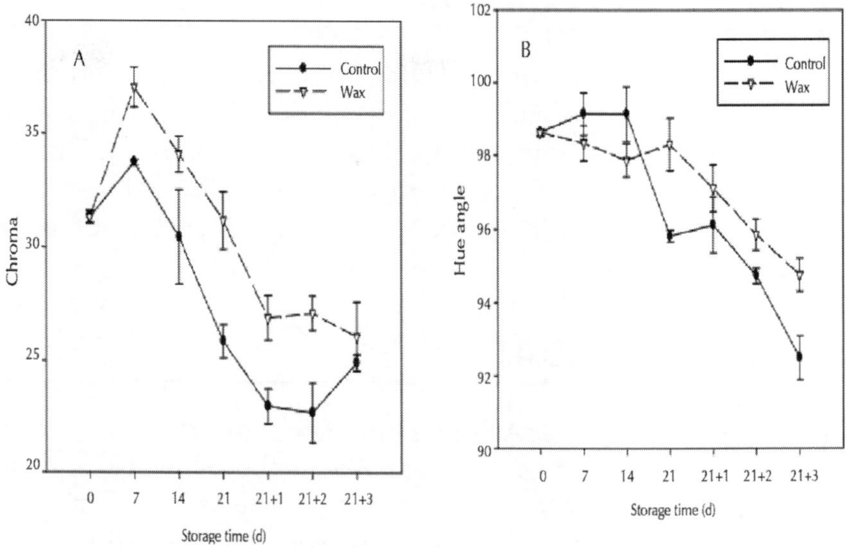

Figure 3: Effects of Sta-Fresh 2952 wax (60 g/l concentration) treatment on chroma (A) and hue angle (B) of pineapple fruits during storage at 7°C for 21 day and then transferred to 25°C for 3 day. "0-21" represent 21 day storage at 7°C and "21+ n" (n= 1, 2, 3) mean 21 day storage at 7°C followed by n day storage at 25°C. Values are the mean of five replicates and error bars represent the standard deviation.

Changes in Soluble Sugars and Titratable Acidity (TA)

As shown in Figure 4a, levels of soluble sugars of pineapple fruits were significantly affected by waxing, with wax-treatment having higher levels of sugars than control during most of days of storage. The soluble sugar content was increased in wax-treatment by 13.94 and 29.32%, respectively at the 21th and the 24th day of storage, when compared with that in control.

For TA, a steady increase was observed in both control and wax-treatment during cold storage time. After 21 day of storage, TA contents of all fruits decreased at the later storage time (Figure 4b). The wax treatment reduced TA by approximately 6 and 5% compared with the control at 14 and 21 day, respectively. These results indicated that wax treatment could reduce TA content of fruits in cold storage.

Changes in Total Soluble Solid (TSS)

As shown in Figure 5, TSS of the control rapidly increased on day 22 and then fell toward the end of storage period. The changes in TSS content were more slowly in wax-treatment than in control. The wax-treatment tends to maintain significantly ($P < 0.05$) lower levels of TSS than compared with that in the control during most days of storage (days 14, 21, 22 and 24).

Ascorbic Acid (ASA)

As shown in Figure 6, the AsA content in the control decreased during the first 14 days and increased gradually in the 3th week and then decreased again in the last three days of storage. The AsA content of wax-treatment decreased during the first 22 days of storage and increased gradually with prolonged storage time. Throughout the storage period, there were significant differences between control and wax-treatment ($P < 0.05$). The AsA content in wax-treatment was 146% higher than that in the control on the 24th day of storage.

Changes in Weight Loss

Fruit weight loss is mainly associated with respiration and moisture evaporation through the skin. Coatings act as barriers, thereby restricting water transfer and protecting fruit skin from mechanical injuries, as well as sealing small wounds and thus, delaying dehydration (Hernandezmunoz et al., 2008). As shown in Figure 7, while the weight losses of both control and wax-treatment increased continuously with storage time ($P < 0.05$), the weight loss of control was significantly greater than that of wax-treatment on the 7th and 14th day of storage. At the end of the storage, the control showed 3.1% loss in weight, whereas the weight losses of the wax-treatment were 2.6%.

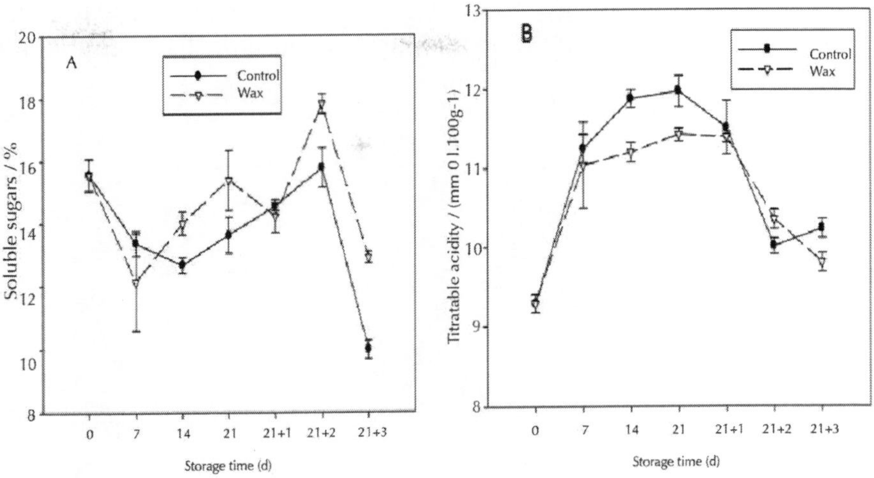

Figure 4: Effects of Sta-Fresh 2952 wax (60 g/l concentration) treatment on soluble sugars (A) and titratable acidity (B) of pineapple fruits during storage at 7°C for 21 day and then transferred to 25°C for 3 day. "0-21" represent 21 day storage at 7°C and "21+ n" (n= 1, 2, 3) mean 21 day storage at 7°C followed by n day storage at 25°C. Values are the mean of five replicates and error bars represent the standard deviation.

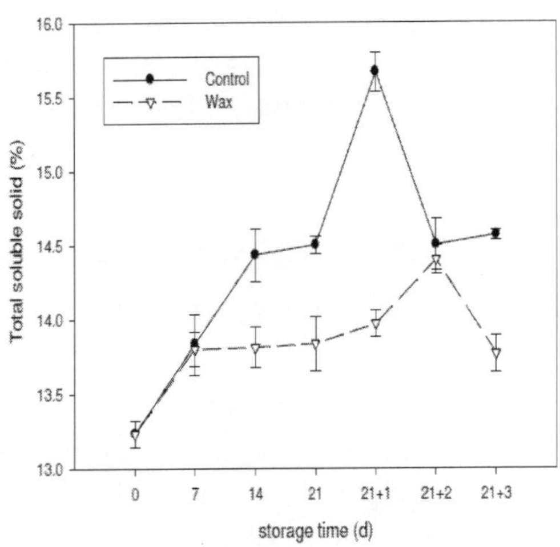

Figure 5: Effects of Sta-Fresh 2952 wax (60 g/l concentration) treatment on total soluble solid of pineapple fruits during storage at 7°C for 21 day and

then transferred to 25°C for 3 day. "0-21" represent 21 day storage at 7°C and "21+ n" (n= 1, 2, 3) mean 21 day storage at 7°C followed by n day storage at 25°C. Values are the mean of five replicates and error bars represent the standard deviation.

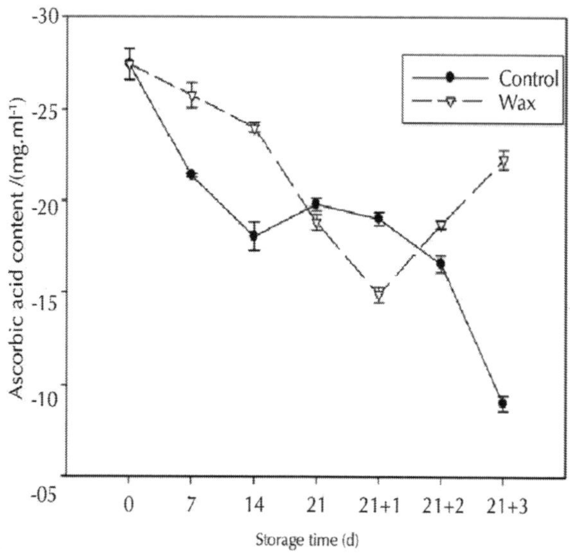

Figure 6: Effects of Sta-Fresh 2952 wax (60 g/l concentration) treatment on ascorbic acid of pineapple fruits during storage at 7°C for 21 day and then transferred to 25°C for 3 day. "0- 21" represent 21 day storage at 7°C and "21+ n" (n= 1, 2, 3) mean 21 day storage at 7°C followed by n day storage at 25°C. Values are the mean of five replicates and error bars represent the standard deviation.

Changes in Relative Leakage Rate and Malondialdehyde (MDA) Content

Relative leakage rate has been used an indicator of injury degree of fruits (Sun et al., 2010). As shown in Figure 8a, the relative leakage rate of both control and waxtreatment was increased continuously during the first 21 days of storage and then decreased continuously during the last 3 days of storage. After 21 days of storage, the relative leakage rate of all samples decreased, which suggested that pineapple fruits might repair the injury of cell membrane by themselves. The wax-treated fruits exhibited significantly ($P < 0.05$) lower levels of relative leakage rate compared with that in the control during the first 23 days of storage.

As shown in Figure 8b, MDA content in both control and wax-treatment showed the same trend of relative leakage rate during the first 21 days of storage and then decreased during the last 3 days of storage. Wax coatings significantly reduced the MDA content of pineapple fruits during the cold storage period compared with the control ($P < 0.05$).

Soluble Protein Content

Soluble protein content in both control and wax-treatment decreased gradually during the whole storage period (Figure 9). The control had a significant decrease in soluble protein content after 21 day storage at 7°C followed by 1 day storage at 25°C, however, the soluble protein content of wax-treatment decreased slightly on the same day (Figure 9). Wax treatment markedly inhibited the decline of soluble proteins, thus, maintaining significantly ($P < 0.05$) higher levels of soluble protein throughout the storage period compared with the control (Figure 9).

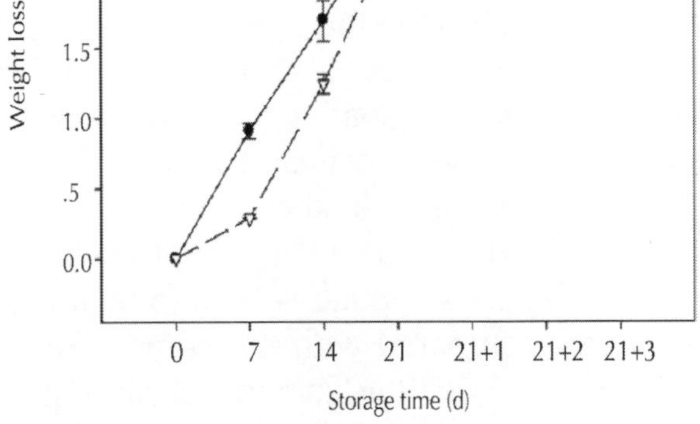

Figure 7: Effects of Sta-Fresh 2952 wax (60 g/l concentration) treatment on percentage weight loss of pineapple fruits during storage at 7°C for 21 day

and then transferred to 25°C for 3 day. "0- 21" represent 21 day storage at 7°C and "21+ n" (n= 1, 2, 3) mean 21 day storage at 7°C followed by n day storage at 25°C. Values are the mean of five replicates and error bars represent the standard deviation.

DISCUSSION

Effects of Different Wax Treatments on Quality of Pineapple Fruits in Cold Storage

The application of wax effectively maintained quality attributes and extended postharvest life of some fruits (Baldwin et al., 1999; Hernandezmunoz et al., 2008; Dang et al., 2010). This study examined the effects of two different types of wax treatment on the quality and physiological changes of harvested pineapple fruits. The results showed that both types of wax treatment can effectively alleviate chilling injury of pineapple fruits and improve their quality after cold storage and it is in agreement with those reported in an earlier study, which has shown that fruits treated with chitosan scored superior quality compared with untreated fruits after cold storage (Wijeratnam et al., 2006). The chilling injury indices of fruits are significantly affected by wax treatment, with wax-treated fruits having a lower index than that of the control. Reuck et al. (2009) have also reported that coating was effective in alleviating chilling injury of litchi fruit. Wax treatment kept firmness of pineapple fruits in a higher level than that in the control and this is similar to that observed by Martinezromero (2005). Fruit acidity and sweetness are two major factors that determine eating taste and quality of pineapple fruits. Wax treatment decreased significantly TA content of fruits, while it increased soluble sugars content after cold storage. This observation is similar to that reported by Sun et al. (2010), who observed that litchi fruits treated with chitosan coating tend to maintain significantly higher levels of sugars compared with the control fruits. Meanwhile, Eum et al. (2009) reported that the coatings of plum fruits slowed the TA changes. Together, these observations indicated that wax treatment not only could alleviate chilling injury, but also improve quality of pineapple in cold storage.

The effects of wax on fruits are associated with the source and concentration of wax, mode of wax application, types of fruits and storage conditions (Hernandezmunoz et al., 2008). The results presented herein indicated that Sta-Fresh 2952 wax treatment (60 g/l concentration) was more effective than the other treatments in alleviating chilling injury index and retarding change of color saturation, the most important factor related to chilling injury in pineapple fruits. Similarly, other studies also observed that the beneficial effects of wax on other types of fruits were enhanced with increasing the wax concentrations (Hernandezmunoz et al., 2008; Zhu et al., 2008).

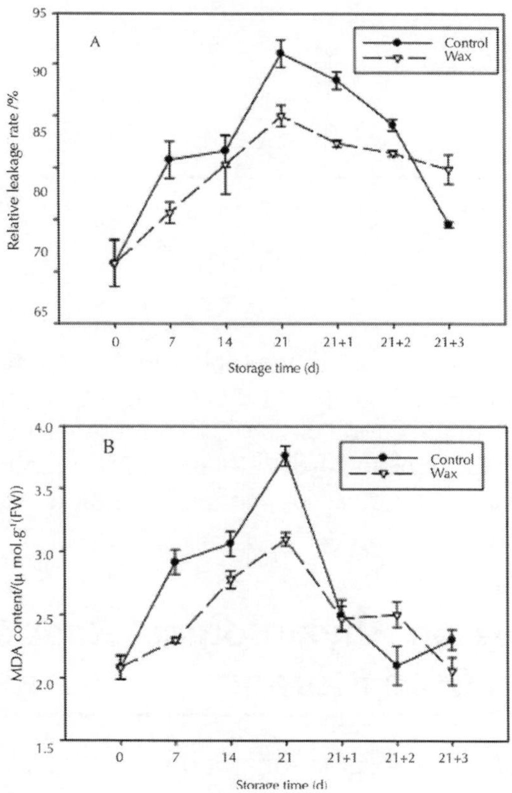

Figure 8: Effects of Sta-Fresh 2952 wax (60 g/l concentration) treatment on relative leakage rate (A) and MDA content (B) of pineapple fruits during storage at 7°C for 21 day and then transferred to 25°C for 3 day. "0-21" represent 21 day storage at 7°C and "21+ n" (n= 1, 2, 3) mean 21 day storage at 7°C followed by n day storage at 25°C. Values are the mean of five replicates and error bars represent the standard deviation.

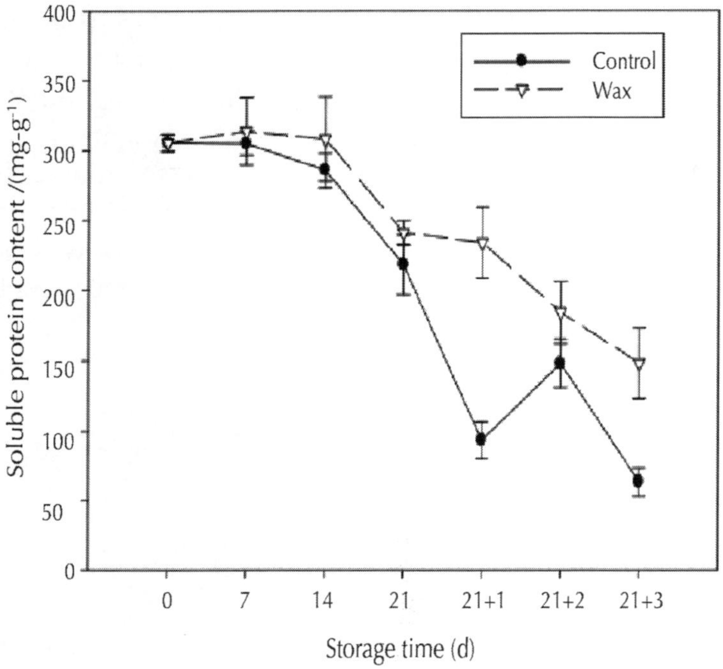

Figure 9: Effects of Sta-Fresh 2952 wax (60 g/l concentration) treatment on soluble protein content of pineapple fruits during storage at 7°C for 21 day and then transferred to 25°C for 3 day. "0-21" represent 21 day storage at 7°C and "21+ n" (n= 1, 2, 3) mean 21 day storage at 7°C followed by n day storage at 25°C. Values are the mean of five replicates and error bars represent the standard deviation.

Effect of Wax on Physiological Responses of Pineapple in Cold Storage

The second important observation made in this study is that wax treatment can maintain the cell membrane integrity of harvested pineapple. Cell membrane is the first barrier that separates cells from their environment and also is a primary target for damage (Cantin et al., 2010). Maintenance of membrane integrity at low temperature has been reported to be important in the resistance to chilling temperature (Antunes and Sfakiotakis, 2008; Wongsheree et al., 2009). It has been reported that the relative electrolyte leakage and MDA content, the proven excellent indicators of cell membrane damage, were gradually increased in fruits with storage time (Cantin et al., 2010; Rui et al., 2010). The results presented herein have shown that the increases in electrolyte leakage and MDA content during cold storage, which indicate the membrane deterioration in pineapple fruits under low temperature stress, were effectively inhibited by wax treatment. Cao et al. (2009) reported that the development of CI in loquat fruits was accompanied by loss of membrane integrity, increases in electrolyte leakage and MDA content. Also, enhanced tolerance to CI by treatment with 1-methylcyclopropene or with hot water was associated with the inhibition of membrane deterioration. The electrolyte leakage and MDA content of all samples decreased gradually after cold storage, suggesting that membrane integrity of pineapple fruits were recovered at room temperature. Similarly, Tsantili et al. (2010) reported that increases in total antioxidant activity that could be at least partially attributed to synthesis of anthocyanins occurred more or less concomitantly with an apparent juice recovery at elevated temperature.

The decrease in AsA level was associated with a reduced capability of preventing oxidative damage and with the incidence of physiological disorders during storage (Lin et al., 2008). Wax treatment kept AsA in a higher level in wax-treat fruits than in control, similar to that reported by Dang et al. (2010), who found that the content of AsA of CA-coated fruits was higher than that of the control. Furthermore, soluble protein levels were higher in wax-treatment than in control. Increasing the soluble proteins in plant could bound more moisture and reduce osmotic stress injury in adversity (Heuer, 2003). This may account for the less weight loss in waxtreatment as compared to the control.

In summary, this study indicates that wax treatment of pineapple fruits could be a potentially useful method to alleviate chilling injury associated with the major changes of cell membrane as indicated by altering electrolyte leakage, MDA content and AsA and to maintain their quality by retarding dehydration, providing a selective barrier to moisture loss and gas exchange, which lead to reduction of general metabolism rate, improvement of textural quality and maintenance of fruit quality. The maintenance of quality and the extension of shelf life of pineapple fruits by waxing during cold storage demonstrated herein suggest that Sta-Fresh 2952 wax could be a potential application as a new effective method for commercial storage and marketing of pineapple fruits and perhaps, other types of fruits.

ACKNOWLEDGEMENTS

This work was funded by the National Natural Science Foundation of China (Project No. UO631004), the Natural Science Foundation of Guangdong Province (Project no. 06200670) and the Ministry of Agriculture of China (Projects no. 3-41 and 2006-G34, A).

REFERENCES

1. Ahmed MJ, Singh Z, Khan AS (2009). Postharvest Aloe vera gelcoating modulates fruit ripening and quality of 'Arctic Snow' nectarine kept in ambient and cold storage. Int. J. Food Sci. Tech., 44: 1024- 1033.
2. Antunes MDC, Sfakiotakis, EM (2008). Changes in fatty acid composition and electrolyte leakage of Hayward' kiwifruit during storage at different temperatures. Food Chem., 110: 891-896.
3. Avallone S, Guiraud JP, Brillouet JM, Teisson C (2003). Enzymatic Browning and Biochemical Alterations in Black Spots of Pineapple [*Ananas comosus* (L.) Merr.]. Curr. Microbiol., 47: 113-118.
4. Baldwin EA, Burns JK, Kazokas W, Brecht JK, Hagenmaier RD, Bender RJ (1999). Effect of two edible coatings with different permeability characteristics on mango (*Mangifera indica* L.) ripening during storage. Postharvest Biol. Technol., 17(3), 215-226.
5. Bradford MM (1976). A rapid and sensitive method for the quantitation of microgram quantities of protein utilizing the principle of protein-dye binding. Anal. Biochem., 72: 248-254.
6. Cantín CM, Crisosto CH, Ogundiwin E A, Gradziel T, Torrents J, Moreno MA. (2010). Chilling injury susceptibility in an intra-specific peach [*Prunus persica* (L.) Batsch] progeny. Postharvest Biol. Technol., 58: 79-87.
7. Cao, S, Zheng Y, Wang K, Rui H, Tang S (2009). Effects of 1-methylcyclopropene on oxidative damage, phospholipases and chilling injury in loquat fruit. J. Sci. Food Agric., 89: 2214-2220. Chen CC, Paull RE (2001). Fruit temperature and crown removal on the occurrence of pineapple fruit translucency. Sci. Hortic-Amsterdam., 88: 85-95.

8. Cheour F, Willemot C, Arul J, Makhlouf DY (1991). Postharvest response of two strawberry cultivars to foliar application of CaCl2. Alexandria, VA, ETATS-UNIS: American Society for Horticultural Science: Hort. Sci., pp. 1186-1188.

9. Chien P, Sheu F, Lin H (2007). Coating citrus (*Murcott tangor*) fruit with low molecular weight chitosan increases postharvest quality and shelf life. Food Chem., 100: 1160-1164.

10. Dang QF, Yan JQ, Li Y, Cheng XJ, Liu CS, Chen XG (2010). Chitosan Acetate as an Active Coating Material and Its Effects on the Storing of *Prunus avium* L. J. Food Sci., 75: S125-S131.

11. Dipierro S, Leonardis DS (1997). The ascorbate system and lipid peroxidation in stored potato (*Solanum tuberosum* L.) tubers. J. Exp. Bot., 48: 779-783.

12. Eum HL, Hwang DK, Linke M, Lee SK, Zude M (2009). Influence of edible coating on quality of plum *(Prunus salicina Lindl.* cv. 'Sapphire'). Eur. Food Res. Technol., 229: 427-434.

13. Fan Y, Xu Y, Wang D, Zhang L, Sun J, Sun L (2009). Effect of alginate coating combined with yeast antagonist on strawberry (Fragaria×ananassa) preservation quality. Postharvest Biol. Technol., 53: 84-90.

14. Hernandezmunoz P, Almenar E, Valle V, Velez D, Gavara R (2008). Effect of chitosan coating combined with postharvest calcium treatment on strawberry (*Fragaria*×ananassa) quality during refrigerated storage. Food Chem., 110: 428-435.

15. Heuer B (2003). Influence of exogenous application of proline and glycinebetaine on growth of salt-stressed tomato plants. Plant Sci., 165: 693-699.

16. Jin P, Zheng Y, Tang S, Rui H, Wang C (2009). A combination of hot air and methyl jasmonate vapor treatment alleviates chilling injury of peach fruit. Postharvest Biol. Technol., 52: 24-29.

17. Ko HL, Campbell PR, Jobin-Décor MP, Eccleston KL, Graham MW, Smith MK (2006). The Introduction of Transgenes to Control Blackheart in Pineapple (*Ananas Comosus* L.) cv. Smooth Cayenne by Microprojectile Bombardment. Euphytica., 150: 387-395.

18. Lin L, Wang BG, Wang M, Cao J, Zhang J, Wu Y, Jiang WB (2008). Effects of a chitosan-based coating with ascorbic acid

on postharvest quality and core browning of 'Yali' pears (*Pyrus bertschneideri Rehd.*). J. Sci. Food Agric., 88: 877-884.

19. Li YH, Wu YJ, Wu B, Zou MH, Zhang Z, Sun GM (2011). Exogenous gibberellic acid increases the fruit weight of 'Comte de Paris' pineapple by enlarging flesh cells without negative effects on fruit quality. Acta Physiologiae Plantarum, pp. 1-8.
20. Martinezromero D, Alburquerque N, Valverde J, Guillen F, Castillo S, Valero D (2006). Postharvest sweet cherry quality and safety maintenance by Aloe vera treatment: A new edible coating. Postharvest Biol. Technol., 39:93-100.
21. Masuko T, Minami A, Iwasaki N, Majima T, Nishimura SI, Lee YC (2005). Carbohydrate analysis by a phenol-sulfuric acid method in microplate format. Anal. Biochem., 339: 69-72.
22. Meng X, Li B, Liu J, Tian S (2008). Physiological responses and quality attributes of table grape fruit to chitosan preharvest spray and postharvest coating during storage. Food Chem., 106: 501-508.
23. Nimitkeatkai H, Srilaong V, Kanlayanarat S (2006). Effect of semi-active modified atmosphere on internal browning of cold stored pineapple. In Purvis AC, McGlasson WB, Kanlayanarat S (Eds.). Proceedings of the IVth International Conference on Managing Quality in Chains, Vols 1 and 2 - The Integrated View On Fruits And Vegetables Quality. Leuven 1: Int. Soc. Hort. Sci., pp. 649-653
24. Nunes MCN, Brecht JK, Morais AMMB, Sargent SA (1995). Physical and chemical quality characteristics of strawberries after storage are reduced by a short delay to cooling. Postharvest Biol. Technol. 6: 17- 28.
25. Paul ER, Rohrbach GK (1985). Symptom development of chilling injury in pineapple fruit (Vol. 110). Alexandria, VA, ETATS-UNIS: J. Am. Soc. Hortic. Sci ., pp. 100-105
26. Qiuping Z, Wenshui X (2007). Effect of 1-methylcyclopropene and/or chitosan coating treatments on storage life and quality maintenance of Indian jujube fruit. Lwt-Food Sci. Technol., 40: 404-411.
27. Reuck DK, Sivakumar D, Korsten L (2009). Effect of integrated application of chitosan coating and modified atmosphere

packaging on overall quality retention in litchi cultivars. J. Sci. Food Agric., 89(5): 915-920.

28. Rui H, Cao S, Shang H, Jin P, Wang K, Zheng Y (2010). Effects of heat treatment on internal browning and membrane fatty acid in loquat fruit in response to chilling stress. J. Sci. Food Agric., 90: 1557-1561.
29. Selvarajah S, Bauchot AD, John P (2001). Internal browning in coldstored pineapples is suppressed by a postharvest application of 1- methylcyclopropene. Postharvest Biol. Technol., 23: 167-170.
30. Soares AG, Trugo LC, Botrel N, Silva SLF (2005). Reduction of internal browning of pineapple fruit (*Ananas comusus* L.) by preharvest soil application of potassium. Postharvest Biol. Technol., 35: 201-207.
31. Soto-Zamora G, Yahia EM, Brecht JK, Gardea A (2005). Effects of postharvest hot air treatments on the quality and antioxidant levels in tomato fruit. Lwt-Food Sci. Technol., 38: 657-663.
32. Sun DQ, Liang GB, Xie JH, Lei XT, Mo YW (2010). Improved preservation effects of litchi fruit by combining chitosan coating with ascorbic acid treatment during postharvest storage. Afr. J. Biotechnol., 9: 3272-3279.
33. Tietel Z, Bar E, Lewinsohn E, Feldmesser E, Fallik E, Porat R (2010). Effects of wax coatings and postharvest storage on sensory quality and aroma volatile composition of 'Mor' mandarins. J. Sci. Food Agric., 90: 995-1007
34. Tsantili E, Shin Y, Nock JF, Watkins CB. (2010). Antioxidant concentrations during chilling injury development in peaches. Postharvest Biol. Technol., 57: 27-34.
35. Tzoumaki MV, Biliaderis CG, Vasilakakis M (2009). Impact of edible coatings and packaging on quality of white asparagus (*Asparagus officinalis*, L.) during cold storage. Food Chem., 117: 55-63.
36. Wijeratnam RSW, Hewajulige IGN, Perera M (2006). The control of black rot disease and the application of a novel wax formulation for extending storage life of pineapple. In Joubert PH (Ed.). Proceedings of the Vth International Pineapple Symposium Leuven 1: Int. Soc. Hort. Sci. pp. 185-189. Hu et al. 7603

37. Wilsonwijeratnam R, Hewajulige I, Abeyratne N (2005). Postharvest hot water treatment for the control of black rot of pineapple. Postharvest Biol. Technol.. 36: 323-327.

38. Wongsheree T, Ketsa S, Doorn VWG. (2009). The relationship between chilling injury and membrane damage in lemon basil (*Ocimum* × citriodourum) leaves. Postharvest Biol. Technol.. 51: 91- 96.

39. Zhang C, Ding Z, Xu X, Wang Q, Qin G, Tian S (2009). Crucial roles of membrane stability and its related proteins in the tolerance of peach fruit to chilling injury. Amino Acids. 39:181-194.

40. Zheng X, Tian S (2006). Effect of oxalic acid on control of postharvest browning of litchi fruit. Food Chem. 96: 519-523.

41. Zhou Y, Dahler JM, Underhill SJR, Wills RBH (2003). Enzymes associated with blackheart development in pineapple fruit. Food Chem., 80: 565-572.

42. Zhu X, Wang QM, Cao JK, Jiang WB (2008). Effects of chitosan coating on postharvest quality of mango (*Mangifera indica* L. cv. Tainong) fruits. J. Food Process. Preserv., 32: 770-784.

Chapter 5

Effect of Storage Periods on Postharvest Quality of Pumpkin

M. A. Rahman[1], M. Miaruddin[2], M. H. H. Khan[1]
M. A. T. Masud[3], and M.M. Begum[4]

[1]Senior Scientific Officer

[2]Principal Scientific Officer, Postharvest Technology Division, Bangladesh Agricultural Research Institute (BARI)

[3]Senior Scientific Officer, Horticulture Research Centre, BARI

[4]Senior Scientific Officer, Tuber Crops Research Centre, BARI, Gazipur-1701, Bangladesh

ABSTRACT

The fully matured pumpkins (Cucurbita moschata Poir) of BARI Pumpkin-1 and BARI Pumpkin-2 were harvested from the experimental field of Horticulture Research Centre, BARI, and Gazipur, Bangladesh. They were then stored under ambient room conditions (27-31 °C and 75-90% RH) for various periods ranged from 15 days to 120 days. The effect of different storage periods was assessed by evaluating their impact on changes in quality attributes of pumpkins. Results indicated that a slow but steady weight loss occurred in pumpkin with maximum loss of 18 and 21% after 120 days of storage in BARI Pumpkin-1 and BARI Pumpkin-2, respectively. β-carotene and ascorbic acid contents were decreased throughout the whole storage time, however, they were drastically reduced during first 30 days of storage. Total soluble solids content increased until 45 days of storage followed by a decrease with progress in time. Titratable acidity was slowly decreased and simultaneously pH value was increased throughout the storage period. The results indicated that storage conditions need to be managed carefully to slow down any changes in fresh pumpkin.

INTRODUCTION

Pumpkin is one of the most important vegetable crops grown extensively throughout the tropical and subtropical countries. Due to its high nutritional content and lucrative market price, pumpkin may be considered as a high value crop. In Bangladesh, both immature and mature fruits are used as a vital ingredient for several culinary preparations. Pumpkins are rich in carbohydrate and minerals and cheaper source of vitamins, especially carotenoid pigments (Bose and Som, 1986), which have a major role in nutrition in the form of pro-vitamin A, antioxidants, when used at ripening stage (Dutta et al., 2006). Thus, this vegetable can contribute to improve nutritional status of the people of Bangladesh, particularly the vulnerable group in respect of vitamin A requirement. Moreover, pumpkin has a long storage capacity compared to other vegetables even in ambient temperature and relative humidity. Generally, the growers of Bangladesh store full ripe pumpkin fruits in their house for 3-4 months under ordinary conditions. Yawalkar (1985) reported that well matured pumpkin fruits can be stored for 2-4 months under normal home conditions and therefore, can meet the demand of vegetables supply during slack season. In a preliminary experiment, it was found that the nutritional value of stored pumpkin fruit get deteriorated during storage at higher temperature (30- 32 °C). During storage, the majority of the physico-chemical changes that occur in the harvested commodities are related with oxidative metabolism, which is highly correlated with change in quality and storage duration (Maiti et al., 2006). Many studies on the postharvest nutrient destruction and textural changes for fruits and vegetables are available, but no study has been made regarding the decline of nutritional quality of pumpkin during storage at ambient conditions. The objective was to study the degradation of β-carotene and other nutrients of pumpkin during storage under ordinary conditions.

MATERIALS AND METHOD

Fruits and Treatments

Fully matured fruits of BARI pumpkin-1 and BARI pumpkin-2 were harvested from the experimental field of Horticulture Research Centre, BARI, Gazipur on 05 May 2010. Twenty seven pumpkin fruits were selected with uniform size, shape and colour from each variety. The surface of the selected fruits were cleaned with soft cloths and kept in storage up to 120 days under ambient room conditions (27-31 °C and 75-90% RH) (Fig.1). Every fortnight, three fruits represented three replications for each variety were used for determining physico-chemical characteristics. A different set of three fruits from each variety was marked to find out the weight loss, and the same set of fruits was used throughout the whole storage period. Data were recorded every fortnight started on the same day of harvest.

Measurements of Weight Loss

To determine weight loss, individual fruit was weighed with a top pan electronic balance (BP 2100, Sartorius, Germany) at the beginning of the experiment and thereafter each fortnight during the whole storage period. Weight loss was expressed as percentage loss of the initial total weight.

Measurement of β-Carotene

The estimation of β-carotene was done by the extraction of 3g fruit sample with acetone (Fisher Scientific Ltd., UK) and petroleum ether. It was further purified with acetone, methanolic KOH and distilled water. The resulting solution was filtered with anhydrous sodium sulphate and read on a spectophotometer (T-80, PG Instrument Ltd., UK) at 451 nm against petroleum ether as a blank. A standard graph was plotted using synthetic crystalline β-carotene (Fluka, Germany) dissolved in petroleum ether and its optical density measured at 451 nm (Alasalvar et al., 2005).

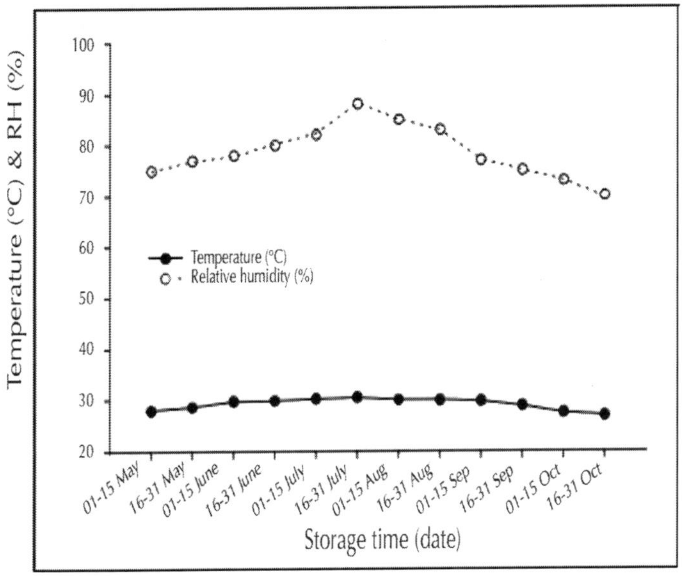

Figure 1: Changes of temperature and relative humidity of the storage room for pumpkin from May to October 2010.

Measurements of Ascorbic Acid, Total Soluble Solids (TSS) pH and Titratable Acidity

Pulp tissues of pumpkin were cut into small pieces. For ascorbic acid measurement, 10 g pulp tissue was immediately homogenized in 50 mL of 3% cold metaphosphoric acid (HPO_3) using a blender for 2 min and filtered through Whatman filter paper No. 2. The clear supernatant was collected for assaying ascorbic acid by 2, 6-dichlorophenolindophenol titration following the method of Ranganna (1986). Ten milliliters of aliquot was titrated with 0.1% 2, 6-dichlorophenolindophenol solution until the filtrate changed to pink colour persisted for at least 15 seconds and the titration volume of 2, and 6-dichlorophenolindophenol was recorded. Prior to titration 2, 6-dichlorophenolindophenol solution was calibrated by ascorbic acid standard solution. Ascorbic acid content was calculated according to the titration volume of 2, 6-dichlorophenolindophenol and results were expressed as mg 100/g fresh weight. Again, 10 grams of pulp tissues was homogenized in 50 mL of distilled water for 2 min using a kitchen blender and filtered through Whatman filter paper No. 2. The supernatant was collected in order to measure total soluble solids using a hand refractometer (Model N-1E, Atago, Japan) and expressed as percentage, pH using a glass electrode pH meter (Delta 320, Mettler, Shanghai) and titratable acidity expressed as citric acid (%) was determined by titration with 0.1 mol/L NaOH to pH 8.1 according to the method by Ranganna (1986).

Experimental Design and Statistical Analysis

The experiment was carried out in a completely randomized design (CRD) with three replications. All the analytical works were repeated three times and mean values were used for analysis. The data were subjected to analysis of variance (ANOVA) using the Crop State Statistical Software version 7.2. The results showing significant differences were then subjected to mean separation using LSD test at $P \leq 0.05$.

RESULTS AND DISCUSSION

Weight Loss

Fruits of the both varieties of BARI Pumpkin-1 and BARI Pumpkin-2 showed a progressive weight loss during 120 days of storage at ambient room temperature (27-31 °C) and relative humidity (75-90%) (Fig. 2). Loss in weight for both varieties increased with increasing storage period. It was observed that BARI pumpkin-2 suffered significantly higher weight loss compared to BARI pumpkin-1 for all sampling time. The losses were 21 and 18% after 120 days of storage for BARI Pumpkin-2 and BARI Pumpkin-1, respectively. Significant differences in weight loss between the pumpkin varieties as well as storage durations were found throughout the whole storage period. Storage

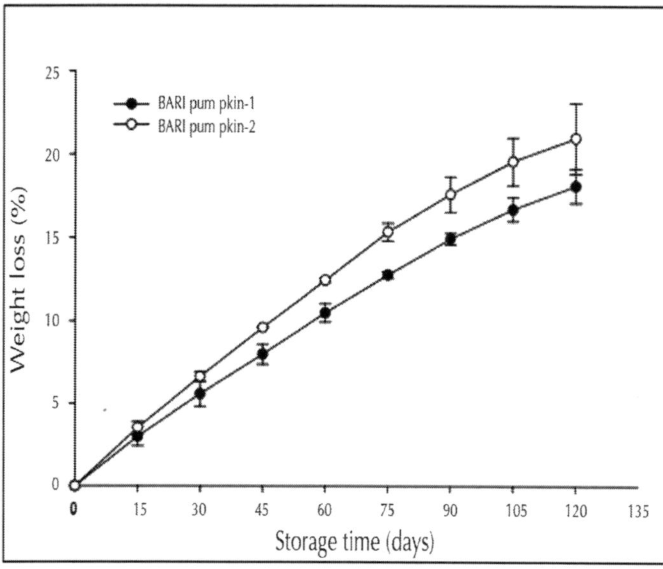

Figure 2: Effect of variety and storage time on the weight loss of pumpkins stored for 120 days under ambient conditions (27-31°C and 75-90% RH). Vertical bar represents standard deviation.

Harvested fruits and vegetables continue to respire and loss water to the environment, which cannot be replaced and weight loss occur (Burdon, 1997; Thompson, 1996). In this experiment, the weight loss of pumpkins occurred at 27-31 °C temperature and 75-90% RH during whole storage period. These results are in agreement with Kumar et al. (1999), who reported that storage duration and temperature having significant effect on weight loss of fresh produces. Possibly, higher rate of transpiration at room temperature (27-31 °C) could be the main cause of higher weight loss. Boxall et al. (2002) also reported that the rate of water loss is largely controlled by the storage temperature and relative humidity.

Total Soluble Solid Content

Changes in the TSS content of pumpkin fruits with storage period is presented in Fig. 3. Different storage periods and pumpkin varieties showed significant variation in the changes of TSS content during whole storage period. The initial TSS contents of BARI Pumpkin-1 and BARI Pumpkin-2 were 6.16 and 6.67%, respectively, which gradually increased until 45 days of storage and reached the maximum levels of 8.1 and 8.3%, respectively. After this period, noticeable decrease in TSS content was recorded for both varieties with the values of 6.6 and 6.8% for BARI Pumpkin-1 and BARI Pumpkin-2, respectively by the end of storage. These findings are in agreement with Maiti et al. (2006), who reported that the general tendency of TSS in climacteric fruits and vegetables is to increase with an initial increase in sugar followed by a decrease.

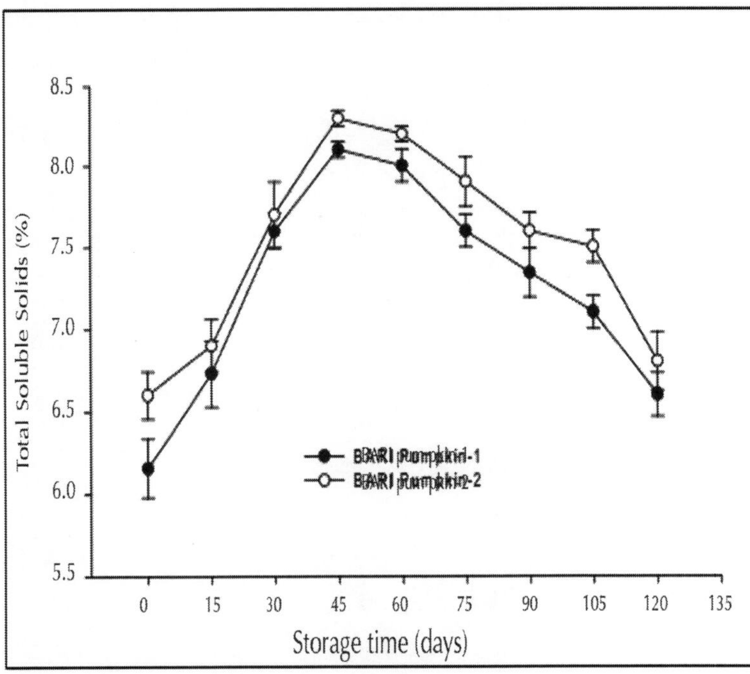

Figure 3: Effect of variety and storage time on the changes in total soluble solids contents of pumpkins stored for 120 days under ambient conditions (27-31°C and 75-90% RH). Vertical bar represents standard deviation.

β-carotene Content

β-carotene degradation of pumpkin fruits which were stored under ambient conditions over a period of 120 days was investigated. It is evident from Fig. 4 that storage duration of pumpkin fruits led to a significant decrease in β-carotene content of both the varieties. The initial β-carotene contents were 36.5 µg/g and 37.3 µg/g for BARI Pumpkin-1 and BARI Pumpkin-2, respectively, which were drastically reduced to 16.5 µg/g and 17.16 µg/g, respectively, after 30 days of storage. These decreases might be possible due to the higher temperature and RH (%) existed in the storage room. In general, oxidation is a major cause of β-carotene destruction (Simpson, 1985). Dutta et al. (2006) reported that the thermal degradation of pumpkin tissue caused in the loss of β- carotene content. Thermal processing also results in losses of trans-β-carotene and formation of cis isomers as reported by many authors (Chen and Chen, 1994; Khachik et al., 1992; Chandler and Schwartz, 1988). After 30 days of storage, pumpkin fruits showed a gradual decline in β-carotene content and exhibited the minimum values of 5.4 µg/g and 5.63 µg/g for BARI Pumpkin-1 and BARI Pumpkin-2, respectively, at the end of storage period.

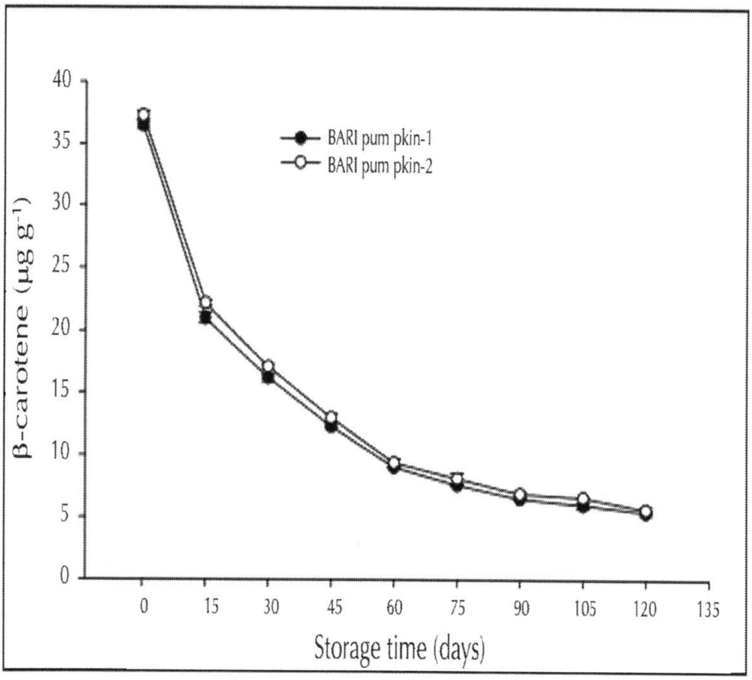

Figure 4: Effect of variety and storage time on the changes in β-carotene contents of pumpkins stored for 120 days under ambient conditions (27-31°C and 75- 90% RH). Vertical bar represents standard deviation.

Ascorbic Acid, Titratable Acidity and pH

Different storage periods induced significant variation in ascorbic acid content of both pumpkin varieties during storage (Fig. 5). Initially, the ascorbic acid contents of BARI Pumpkin-1 and BARI Pumpkin-2 were recorded by 11.2 and 9.31 mg/100g, respectively. The contents were gradually declined over the storage period and reached the minimum levels of 5.2 and 4.91 mg/100g for BARI Pumpkin-1 and BARI Pumpkin-2, respectively, by the end of storage. Both the varieties and different storage time showed significant differences in ascorbic acid contents during whole storage period. Titratable acidity (TA) and pH of pumpkins were also significantly affected by storage durations. At the beginning of storage (day '0'), the TA of pumpkins was higher with the values of 0.17 and 0.15% for BARI Pumpkin-1 and BARI Pumpkin-2, respectively (Fig. 6). As storage progressed TA was slowly declined and reached the minimum levels of 0.11% for both varieties by the end of storage. The changes in pH of pumpkins as function of varieties and different storage periods are shown in Fig. 7. The initial pH values were 6.49 and 6.47 for BARI Pumpkin-1 and BARI Pumpkin-2, respectively, which gradually increased during storage with significant differences ($P \leq 0.05$) between varieties and storage periods. At the end of storage, pH values reached the maximum levels of 6.85 and 6.80 for BARI Pumpkin-1 and BARI Pumpkin-2, respectively

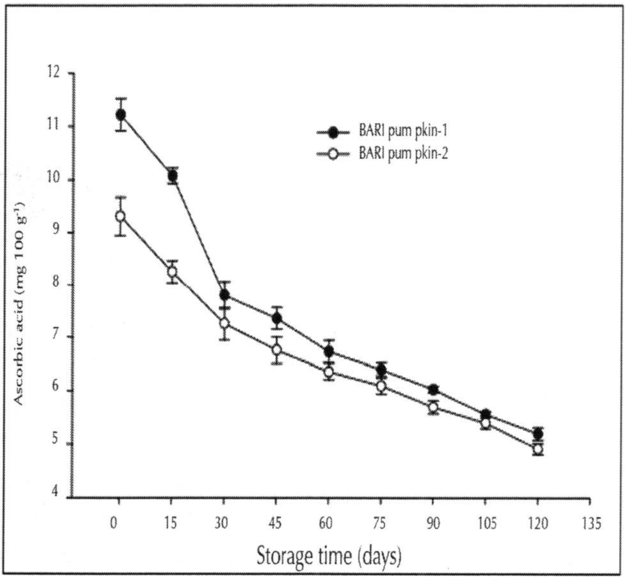

Figure 5: Effect of variety and storage time on the changes in ascorbic acid contents of pumpkins stored for 120 days under ambient conditions (27-31°C and 75-90% RH). Vertical bar represents standard deviation.

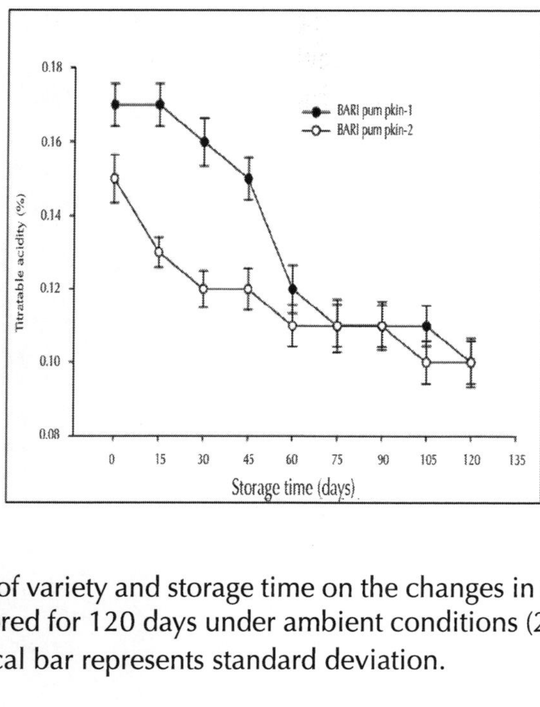

Figure 6: Effect of variety and storage time on the changes in titratable acidity of pumpkins stored for 120 days under ambient conditions (27-31°C and 75-90% RH). Vertical bar represents standard deviation.

The changes in ascorbic acid contents and titratable acidity during storage vary depending on the crop maturity and storage temperature (Maiti et al., 2006). The recommended storage temperature and RH of pumpkin are 12.2 °C and 70- 75%, respectively, for 84-160 days (SeaLand, 1991). In another experiment, Snowdon (1991) found that the best temperature and RH for pumpkin were 10-13 °C and 60-70%, respectively, for 2-5 months of storage. In the present study, pumpkin fruits were stored at 27-31 °C and 75-90% RH (ambient room conditions), which were much higher than that of recommended storage conditions. The higher temperature and RH prevailed in the storage room might be the major causes for rapid declining of ascorbic acid and TA content. The results are in agreement with Maiti et al. (2006) who reported that the ascorbic acid contents decrease rapidly at higher storage temperature. As acidity of the fruits decreased, simultaneously the pH values increased over the storage period.

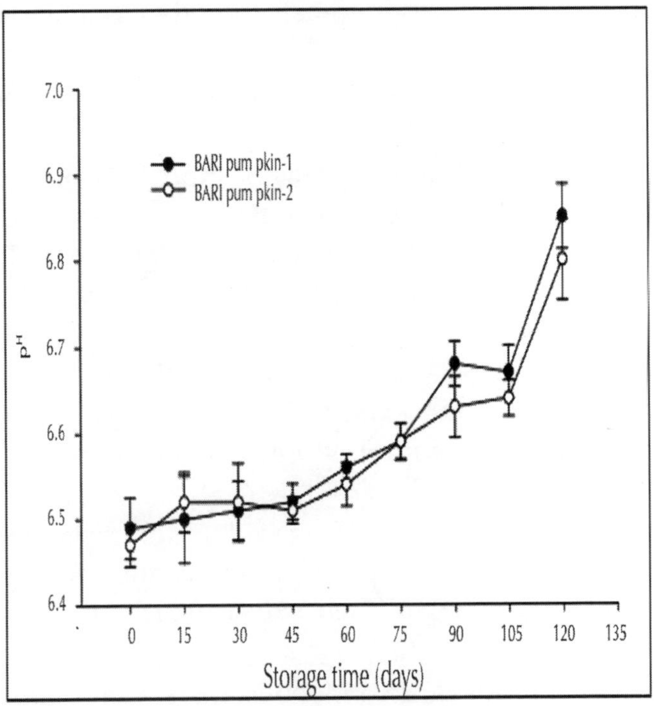

Figure 7: Effect of variety and storage time on the changes in pH of pumpkins stored for 120 days under ambient conditions (27-31°C and 75-90% RH). Vertical bar represents standard deviation.

Results of the present study showed that long term storage of pumpkin at ambient room temperature and RH led to a considerable degradation of nutritional values of pumpkin particularly in β-carotene, ascorbic acid, and TSS contents. Both storage duration and pumpkin varieties had significant effect on the degradation of these nutritional elements. The rate of degradation of β- carotene and ascorbic acid was much higher during first 30 days of storage. Further studies are required to optimize the storage conditions of pumpkins for maintaining better fruit quality during storage.

REFERENCES

1. Alasalvar, J. E., M. Al-Farsi, P.C. Quantic, F. Shahidi, and R.Wiktorowicz. 2005. Effect of chill storage and modified atmosphere packaging (MAP) on antioxidant, phenolics and sensory quality of ready-to-eat shredded orange and purple carrots. Food Chem. 89: 69-76.
2. Bose, T.K. and M.G. Som. 1986. Vegetable Crops in India. Naya Prokash, Calcutta. Pp. 92-95.
3. Boxall, R.A., J.R. Brice, S.J. Taylor, and R.D. Bancroft. 2002. Technology and management of storage. In: Crop Post-Harvest: Science and Technology-Principles and Practice, Vol. 1. Golob, P., Farrell, G. and Orchard, J.E. (eds). Blackwell Publishing, Pp. 141-232.
4. Burdon, J.N. 1997. Postharvest handling of tropical and subtropical fruits for export. In: Postharvest Physiology and Storage of Tropical and Subtropical Fruits. Mitra, S. (Ed), CAB International, UK. pp. 01-19.
5. Chandler, L.A. and S.J. Schwartz. 1988. Isomerization and losses of trans β-carotene in sweet potatoes as affected by processing treatment. J. Agric. Food Chem. 36: 129- 133.
6. Chen, T.M. and B.H. Chen. 1994. Optimization of mobile phase for the HPLC of cistrans carotene isomers. Chromatographia 39: 346-354.
7. Dutta, D., U.R. Chaudhuri, and R. Chakraborty. 2006. Effect of thermal treatment on the β-carotene content, colour and textural properties of pumpkin. J. Food Sci. Technol. 43 (6): 607-611.

8. Kahchik, F., M. Goli, G. Beecher, J. Holden, W. Lusdy, M. Tenorio, and M. Barreia. 1992. Effect of food preparation on qualitative and quantitative distribution of major carotenoid constituents of tomatoes and several green vegetables. J. Agric. Food Chem. 40: 390-398.
9. Kumar, A., B.S. Ghuman, and A.K. Gupta. 1999. Non-refrigerated storage of tomatoeseffect of HDPE film rapping. J. Food Sci. Technol. 36: 438-440.
10. Maiti, R.K., N.C. Sarkar, and V.P. Singh. 2006. Principles of Post-harvest Seed Physiology and Technology. Agrobios, India. p. 450.
11. SeaLand. 1991. Shipping guide to perishables. SeaLand Service Inc., PO Box 800, Iselim, New Jersey 08830, USA.
12. Simpson, K.L. 1985. Chemical changes in natural food pigments. In: Chemical changes in food during processing. Richardson, T. and J.W. Finley (eds), AVI. Westport, CT, Pp. 409-441.
13. Snowdon, A.L. 1991. A Colour Atlas of Postharvest Diseases and Disorders of Fruits and Vegetables. Vol. 2. Wolfe Scientific Ltd., London, P. 416.
14. Ranganna, S., 1986. Handbook of Analysis and Quality Control for Fruit and Vegetable Products, 2nd Edition. Tata McGraw-Hill, New Delhi, India, P. 1112.
15. Thompson, A.K. 1996. Postharvest Technology of Fruits and Vegetables. Blackwell Science Ltd. P. 410
16. Yawallkar, K.S. 1985. Vegetable Crops in India, 3rd Edition, Agric. Hort. Pub. House, Nagpur-440010. Pp. 158-162.

Chapter 6

The Concepts and Problems of Post–harvest Food Losses in Perishable Crops

Atanda S A[1], Pessu P. O[1], Agoda S[1], Isong I. U, and Ikotun I[2]

[1]Nigerian Stored Products Research Institute (NSPRI), 32/34 Barikisu Iyede Street, Yaba, P. M. B. 12543 Lagos, Lagos State, Nigeria.

[2]Nigerian Stored Products Research Institute behind Red Cross office, Onireke, P. M. B. 5044 Ibadan, Oyo State, Nigeria.

ABSTRACT

The loss of foods in the post-harvest system is not new; it has always been a problem for mankind. In these days of rapidly enlarging population in the poorest countries of the world where food is already short, there is an increasing urgency to do a better job of conserving mankind food supply in order to alleviate hunger and malnutrition. The main objective of this review is to find out the concepts and problems of post-harvest food losses in perishable crops. It examined the various concepts of postharvest food losses, the importance of perishable crops, causes of food losses, environmental consideration and its influence on food losses. It also sought solutions to some of the identified problems. It was established from the review that the factors contributing to these food losses include; the initial quality of the crop, mechanical injury, temperature, storage atmosphere, genetic factors and environmental influence. In order to minimize these problems, the appropriate agricultural techniques such as the general principles of extending shelf-life of these crops must be put in place. There should be proper management of temperature, humidity and effective methods for preventing these losses. Since most national governments acknowledge that post – harvest food losses is complex, therefore, it requires a commitment to an integrated approach, involving numerous organizations, including local communities and groups.

INTRODUCTION

Most studies on post-harvest technology have so far concentrated on grains and other durable products, which are stored dry and a substantial technology has been developed to deal with these problems. Less work has been undertaken on the perishable food crops, yet they are of great importance in many parts of the humid and sub-humid tropics and contribute the staple carbohydrate portion of the diets of some 500 to 700 million people in the developing countries (Lancaster and Coursey, 1984). In the case of the tropical perishable staple foods, which have no close analogues in temperate zone agriculture, this neglect of the traditional wisdom is especially unfortunate, as the underlying philosophies of the cultures in which they are extensively grown are not neglected [or expressed as so much pre-Galilean as nonGalilean] and are extremely alien to those of Europe, within which scientific thinking developed (Coursey, 1976, 1978a).

These perishable staple foods are very largely produced from small-scale subsistence level systems and the technologies employed in both production and utilization is usually simple and founded on longestablished traditional practice. The most important are the root; cassava, yams, the various aroids, sweet potatoes and white potatoes (Coursey and Haynes, 1970), the total production of which is now around 185 million t/a; fruits such as cooking bananas, (plantains) and breadfruit are also important, still the former being a major food with a world production of over 20 million t/a, while a proportion of the dessert banana crop also iseaten as a staple, cooked while unripe (Burden and Coursey, 1977); there are also crop products derived from vegetative organs such as stem starches of palms and other types of plant.

There is a war going on that began millions of years ago. Although, the many generations of soldiers have not changed a great deal in appearance during this time, the tactics and weapons have grown more sophisticated. Each army has won a share of the battles, but the consummate victory has eluded both. Neither side can afford to give up' for nothing less than the sustenance of life is at stake. "The war that is referred to is, of course, the war between humankind and certain species of insects, weeds, pathogens, nematodes, rodents and other pest that daily compete with our crops, gnaw at our dwellings, infest our domestic animals or destroy our health" (Kuhr, 1979).

The statement in the foregoing vividly describes the problem humans always face to preserve food supply after they have produced it. The loss of foods in the postharvest system is not new; it has always been a problem for mankind. In these days of rapidly enlarging populations in the poorest countries where food is already short, there is an increasing urgency to do a better job of conserving mankind's food supply in order to alleviate hunger and malnutrition. Some far-sighted individuals have been drawing attention to the problem of post-harvest losses for many years.

However, fresh horticultural produce is highly perishable with some estimates suggesting a postharvest loss of 30 to 50% in fruits and vegetables. The loss occurs due to poor pre-production and post-harvest management as well as lack of appropriate processing and marketing facilities. These losses have several adverse impacts on farmer income, consumer prices and nutritional quality of the produce. Because of the poor planting material, cultural practices including harvesting methods and handling practices, the quality of harvested produce is below standard. Absence of farm storage facility and proper pack house/packing station results in the perishable produce being marketed immediately after harvesting without primary processing and adequate packaging. The solid wastes originating from horticultural crops in metro cities, can create drainage problems and cause water logging, as well as invite stray animals near garbage dumps. These bio-wastes also deteriorate very rapidly causing unhygienic conditions, increasing atmospheric pollution and provide a breeding ground for pests (Expert Consultation, 2010).

The United Nations General Assembly resolution of September 1975 focused world-wide attention on the problem of post-harvest food losses and called for concerted action to reduce these losses in the following words: "The further reduction of post-harvest food losses in developing countries should be undertaken as a matter of priority' with a view to reaching at least a 50% reduction by 1985. All countries and competent international organizations should co-operate financially and technically in the effort to achieve this objective". As a result of this resolution, a number of national and international donor agencies expanded existing programmes and initiated new programmes that were directed to the problem of reducing post-harvest losses. Most of these are grain legumes.

A report prepared by the United States National Academy of Science in 1978 on the problem of postharvest food loaves in developing countries pointed out the need for giving consideration to losses in food products other than the cereals, particularly roots and tubers, fruits and vegetables. Largely as a result of this report, the donor organizations are beginning to consider intervention programmes that can reduce losses in horticultural crops.

Post-harvest losses of fruits and vegetables are more serious in developing countries than those in well developed countries. An additional constraint to improving this situation is that in most developing countries the number of scientists concerned with postharvest food losses is significantly lower than those involved in production research. In the early days of horticulture in wolf developed countries, heavy losses occurred in much the same manner as they do today in developing countries. Increasing industrialization in technologically advanced nations gradually brought about improvements in crop handling. Elaborate harvesting equipment replaced the crude harvesting tools. Collection centres were strategically established in major producing areas. Containers were remodelled to add more protection to the produce. Commercial storage plants were installed and grade standards adopted. Engineers and economists became more and more aware of raw material behaviour. Concomitant advances in refrigeration technology in the developed countries have made possible establishment of cold chains for the entire postharvest and handling operations. At the institutional level post-harvest research was initiated. Pilot packing houses were installed, coupled with the development of intensive training programmes, the improvement of product quality and reduction in post-harvest losses became the main concern of producers, middlemen, marketing specialists and consumers. Today, enormous volumes of quality horticultural crops produced in technologically advanced countries are made available to millions of people through improved post-harvest handling. Thus, historically and by necessity, post-harvest technology is part of the normal development processes in agriculture. These handling procedures are not fully recognized in less developed countries. Here agriculture may be characterized as disjointed. Production is not linked with marketing. With perishable crops like fruits and vegetables, storage, packaging, transport and handling technologies are practically non-oxiatant; hence, a (FAO, 1981) considerable amount is used to produce loaf. Thus, as more fresh fruits and vegetables are needed to supply the growing population in less developed countries, as more produce is transported to non-producing areas, and as more commodities are stored longer to obtain a yearround supply, post-harvest lose prevention technology measures become paramount. It is distressing to note that so much time is being devoted to the culture of the plant, so much money spent on irrigation, fertilization and crop protection measures only to be wasted about a week after harvest. It is, therefore, important that postharvest procedures should be given much attention as production practices, and there must be a mutual undertaking, in the stages from planting until the product reached the consuming public, between the growers and those who will handle the products after harvest. Fruits and vegetables, roots and tubers (horticultural crops) are quite different in nature from cereal grains and oilseeds, the differences are summarized in Table 1. The causes of spoilage, the rate at which spoilage occurs, the degree of spoilage and the actions needed to reduce spoilage are substantially different for cereals. Because of these differences it is necessary to design a different set of intervention programmes to reduce post-harvest losses in horticultural products. Typical losses range from 10 to 25%, but it is estimated that at least 5% and as much as 100% of a given crop can be lost between the field and the consumer (NAS, 1978; Coursey, 1983). Typically, the main challenge has been to produce enough food to feed the growing population and making it accessible to people of all categories. This requires a well devised protocol to transform production policies to deliver proper market impact that would foster food accessibility as well as income of all players along all concerned value chains. It would however, make bigger impact if there are proper technologies appropriate to local conditions that enables small-scale farmers employed in agriculture to reap from farming activities. The question however, is the type of technology, the capacity of farmers to understand and implement the technology as well as appropriate time for dissemination (Nyamulinda et al., 2011).

Table 1: Comparison of horiticultural crops

Cereals and oilseeds	Horticultural crops
Low moisture content, typically 10 to 20%.	High moisture contort, typically 70 to 95%.
Small unit size, typically less than 1 g.	Large unit size, typically 5 e to 5 kg.
Very low respiration rate with very small generation of heat.	High to very high respiration rate.
Heat production is typically 0.05 mega joule/ton/day for dry grain.	Heat production is typically 0.5 to 10 mega joule/ton/day at 0°C to 5 to 70 mega joule/ton/day at 20°C.
Hard texture.	Soft texture, easily bruised
Stable, natural shelf life is one to several years	Perishable, natural shelf life is a few days to several months.
Losses usually caused by molds, insects and rodents sprouting and bruising.	Losses usually caused by rotting (bacteria, fungi), senescence.
Losses in LDCs usually 10 to 20%.	Losses in LDCs usually 15 to 50%

Source: (FAO 1981).

IMPORTANCE OF PERISHABLE CROPS

Horticultural crops are essential for a nutritionally balanced diet. Fruits and vegetables are the major source of vitamins A and C, a good source of calcium and iron and they supply part of the requirements for a number of other minor nutrients. Roots, tubers, bananas and plantains are important sources of calories and also supply a number of minor nutrients and some protein. In addition, horticultural crops add variety, enjoyment and a sense of satisfaction with the diet because of their appealing colours, flavours and textures. For example, it has been said that although onions and garlic are not rich in nutrients, they make a vegetarian diet acceptable because of the savoury flavour they impart to the monotonous starchy diet in a developing country. On all of these counts-economics, nutrition, acceptabilityhorticultural crops play a major role in developing countries, amply justifying the contortion that something should be done to reduce the high losses that presently occur in these commodities (Tables 2 and 3).

CAUSES OF FOOD LOSSES IN PERISHABLE CROPS

There are so many causes for losses in the post-harvest food chain that it helps to classify them into 2 groups and a number of sub-groups.

Table 2: Major production of horticultural crops in developed countries and developing countries

Commodity	Production 1978 (million tone)		Percentage of cereal production	
	World	Developing countries	World	Developing countries
Cereals	1,580.8	730.6	100%	100%
Roots and tubers	522.9	290.1	33.1	39.7
Vegetables and melons	327.2	189.2	20.7	25.9
Fruits	261.9	141.6	16.6	19.4

Total horticultural production	1,112.0	620.9	70.4	85.0

Source: (FAO Production Yearbook 1979).

Table 3: Estimates of post–harvest losses in perishable staples (percent)

Commodity	Early TPI estimates[1]	NAS[2] (1978) estimates
Potatoes	8. 30	5 to 40
Sweet Potatoes	35 to 65. 95 55 to	55 to 95
Yam	5. 15	10 to 60
Cassava	-	10
Taro	12 to 15	-
Plantains	33	35 to 100

Source: (Coursey, 1983; Coursey and Proctor 1975; Coursey and Booth, 1977).

Primary Causes of Loss

These are the causes that directly affect the food. They may be classified into the following groups:

Biological

Consumption of food by rodents, birds, monkeys and other large animals causes direct disappearance of food. Sometimes the level of contamination of food by the excreta, hair and feathers of animals and birds is so high that the food is condemned for human consumption. Insects cause both weight losses through consumption of the food and quality losses because of their frass, webbing, excreta, heating, and unpleasant odours that they can impart to food.

Microbiological

Microorganisms cause damage to stored foods (e.g., fungi and bacteria). Micro-organisms usually directly consume small amount of the food but they damage the food to the point that it becomes unacceptable because of rotting or other defects. Toxic substances elaborated by molds (known as mycotoxins), cause some food to be condemned and hence lost. The best known mycotoxins is aflatoxin (a liver carcinogen), which is produced by the mold *Aspergillus flavus*. Another mycotoxin which is found in some processed apple and pear products is patulin, which is formed in the apple by rotting organisms such as *Penicillium expansum* which infect fresh apples before they are processed.

Chemical

Many of the chemical constituents naturally present in stored foods spontaneously react causing loses of colour, flavour, texture and nutritional value. An example is the Maillard relation' that causes browning and dicolouration in dried fruits and other product, There can also be accidental or deliberate contamination of food with harmful chemicals such as pesticides or obnoxious chemicals such as lubricating oil.

Biochemical Reactions

A number of enzyme-activated reactions can occur in foods in storage giving rise to oft-flavours, discolouration and softening. One example of this problem is the unpleasant flavours that develop in frozen vegetables that have not been blanched to inactivate these enzymes before freezing.

Mechanical

Bruising, cutting' excessive pooling or trimming of horticultural products are causes of loss.

Physical

Excessive or insufficient heat or cold can spoil foods. Improper atmosphere in closely confined storage at times causes losses.

Physiological

Natural respiratory losses which occur in all living organisms account for a significant level of weight lose and moreover, the process generates heat. Changes which occur during ripening, senescence, including wilting and termination of dormancy (e.g., sprouting) may increase the susceptibility of the commodity to mechanical damage or infection by pathogens. A reduction in nutritional level and consumer acceptance may also arise with these changes. Production of ethylene results in premature ripening of certain crops.

Psychological

Human aversion, such as "I don't fancy eating that today" In some cases food will not be eaten because of religious taboos. Microbiological, mechanical and physiological factors cause moat of the losses in perishable crops.

Secondary Causes of Loss

Secondary causes of loss are those that load to conditions that encourage a primary cause of loss. They are usually the result of inadequate or non-assistant capital expenditures, technology and quality control. Some examples are:

- Inadequate harvesting, packaging and handling skills
- Lack of adequate containers for the transport and handling of perishables.
- Storage facilities inadequate to protect the food.
- Transportation inadequate to move the food to market before it spoils.
- Inadequate refrigerated storage.
- Inadequate drying equipment or poor drying season.
- Traditional processing and marketing systems can be responsible for high losses.
- Legal standards can affect the retention or rejection of food for human use by being too lax or unduly strict.

- Conscientious, knowledgeable management is essential for maintaining tool in good condition during marketing and storage.
- Bumper crops can overload the post-harvest handling system or exceed the consumption need and cause excessive wastage.

ENVIRONMENTAL CONSIDERATIONS AND ITS INFLEUNCE ON FOOD LOSSES

There is increasing recognition that managing the transition to a sustainable food security should become the new central organizing principle. Towards this end, government must take steps to modify existing policies Atanda et al. 607 and provide direction in the formulation of food security policies. Hence, the United Nations environment programme (UNEP) is interested in promoting the health and well-being of both people and the environment, as well as sustainable development. Reducing food losses should improve nutritional status and human health especially in those countries where a large population is inadequately fed. Decreasing food losses offers an opportunity to reduce the pressure on the land and still deliver the same quantity of food to the table, thus reducing to some extent environmental damage caused by agricultural practices. Hence, UNEP encourages the conservation of food because of its positive environmental effects. UNEP is interested in efficient and non–wasteful utilization of resources. UNEP activities include reviewing proposed new initiatives to determine the environmental impact of these initiatives before they are put into effect and to help select from among competing initiatives that are most desirable from the environment stand point to encourage their use and to do this in the early planning stages of the project. Hence, UNEP was concerned about the consequences of manmade chemicals that are used and discharged into the environment with resultant effect on post – harvest food losses. The environmental approach in this case is not to oppose the use of chemicals but to point out the need for using them properly and carefully. The United Nations environment programme supports and promotes ecologically sound and sustainable development. Food loss reduction is an important activity in which UNEP has an interest because this will increase the resource base as well as enhance the environment.

The perishable crops, which for reasons of their importance to human nutrition, their magnitude of production and their vulnerability to spoilage, have common characteristics and problems. To this end the United Nations environment programme and F.A.O organized a 4–day meeting in May, 1980 at FAO headquarters in Rome with 15 specialists from the various post–harvest disciplines and with broad geographical coverage to examine the present status of the post–harvest food losses occurring in perishable crops and the means to reduce these losses.

The Major Environmental Influences on the Keeping Quality of Food are: *Temperature*

In general, the higher the temperature the shorter the storage life of horticultural products and the greater the amount of loss within a given time, as most factors that destroy the produce or lower its quality occur at a faster rate as the temperature increases; this means that the rate of growth of spoilage micro-organisms, the rate of indigenous physiological change and physical processes such as water loss and wilting are affected.

Humidity

There is movement of water vapour between food and its surrounding atmosphere in the direction towards equilibrium water activity in the food and the atmosphere, a moist food will give up moisture to the air while a dry food will absorb moisture from the air. Fresh horticultural products have high moisture content and need to be stored under conditions of high relative moisture loss and wilting, exceptions of onions and garlic. Dried or dehydrated products need to be stored under conditions of low relative humidity in order to avoid absorbing moisture to the point where mold growth occurs.

Solar Radiation

The solar radiation that falls upon foods held in direct sunlight increases the temperature above the ambient temperature. The amount of increase in temperature depends on the intensity of the radiation, the size and shape of the food and duration of exposure to the direct rays of the sun. The intensity of radiation is greater in tropical zones than in temperate zones. Where solar radiation is much higher, considerable quantities of food can often be found in the direct rays of the sun deteriorating in quality at a rapid rate.

Altitude

Research findings indicate that the prevailing temperature is dependent upon the elevation when other factors are equal. There is on the average a drop in temperature of 6.5°C for each kilometre increase in elevation above sea level. Storing food at high altitudes will therefore tend to increase the storage life and decrease the losses in food provided it is kept out of direct rays of the sun.

Atmosphere

The normal atmosphere contain by volume, approximately 78% nitrogen, 21% oxygen, 1% argon, 0.03% carbon dioxide, various amount of water vapour and traces of inert gases. Modifying the atmosphere can improve the shelf and reduce wastage of certain foods. One type of controlled atmosphere storage (CA) is refrigerated storage in which the level of oxygen is reduced to about 3per cent with the carbon dioxide content being raised to 1 to 5%, depending on the commodity. This CA storage may double the storage life over that of regular cold storage for certain varieties of apples and pears by showing down the natural rate of respiration. Many fruits generate ethylene gas during ripening and the presence of this gas accelerates the rate of ripening. If the ethylene is removed from the atmosphere surrounding these fruits as it is generated, their storage life may be extended. Experiments have shown that placing such fruits in a fairly gas–tight container with potassium permanganate, which absorbs ethylene gas, can substantially extend the storage life even at ambient temperature.

Time

The longer the time the food is stored the greater is the deterioration in quality and the greater is the chance of damage and loss. Hence, storage time is a critical factor in loss of foods especially those that have a short natural shelf life.

Biological Pressure

Bacteria and fungi are always present in the atmosphere to contaminate food and cause spoilage should conditions favour their growth. However, it should be emphasized that the contamination or inoculation process with bacteria and fungi occurs to an equal extent during the harvesting process. Soil organisms as well as foliage pathogens can be introduced. Bacteria that cause disease in plants are not usually introduced from the air except in aerosols. Micro–organisms can multiply very rapidly whenever conditions are favourable for growth. The only foods that are free from micro–organism are those that have been thermally processed such as can goods. A similar situation occurs with insects. Insects are in the field and can accompany foods as they are brought from the field into storage, stored food insects are ubiquitous, hiding in storage facilities and moving with store foods when they are moved. Rodents and sometimes birds can exert biological pressure similar to those of insects and micro – organisms.

WAYS OF REDUCING POST–HARVEST FOOD LOSSES IN PERISHABLE CROPS

According to Oluwole Olatunji, the then Director General, Federal Institute of Industrial Research, Oshodi, Lagos during the FIIRO technological open day 2008, entitled "Food security issue in Nigeria: FIIRO's relevance" He said that Nigeria is the fifth on the hunger index rating in the world and the ninth largest population in the world with a very high level of poverty. He therefore enjoined the federal government to establish more food storage facilities in the country to reduce postharvest losses and reduce soaring food prices. He lamented that the nation's food storage capacity was low, resulting in post-harvest loss of above 30% of national agricultural output. Therefore, the achievement of self-food sufficiency aspiration in perishable crops production of the nation, now and future can only be obtained by reducing post-harvest food losses through the appropriate agricultural techniques such as the general principles of extending shelf life of these crops.

These principles will be considered under the following heading: varietal differences, harvesting, handling, sorting, packaging, transportation, storage and processing.

Varietal Differences in Storability

Some varieties of the same crop store better than others. Therefore, to reduce food loss and to achieve maximum shelf-life, only varieties known to store well should be stored.

Harvesting

Harvesting should be carried out as carefully as possible to minimize mechanical injury such as scratches, punctures and bruises to the crop. This should be carried out during the cool part of the day, which is early morning and late evening harvest crop at a matured green state. Immature fruits may not ripen and fruits which are already ripe will have short storage life.

Handling

Mechanical injury provides sites for pest attack and increases physiological losses. Therefore, avoid mechanical injury to the crop while handling. Because of their soft texture, all horticultural products should be handled gently to minimize bruising and breaking of the skin. Bruising renders the product unsaleable to most

People although it usually has minor effect upon the nutritional value. The skin of horticultural products is an effective barrier to most of the opportunistic bacteria and fungi that cause rotting of the tissues. Breaking of the skin also stimulates physiological deterioration and dehydration. Careful digging and movement of roots and tubers significantly reduces post-harvest losses. Careful handling of fruits and vegetables to minimize bruising and breaking of the skin likewise is a well-known method of reducing post-harvest losses as is the provision of adequate shipping containers to protect the produce from bruising' and puncturing of the skin. Reducing the number of times the commodity is handled reduces the extent of mechanical damage. Crops should not be heaped at collection centre on the farm as this will lead to rapid spoilage particularly of fruits at centre of the heap. Fruits and vegetables should not be exposed to the sun. Crops should be transferred promptly after harvest to a clean, cool, well-ventilated shed.

Sorting and Cleaning

Sort out damage crops from undamaged crops. Remove dirt from sound crop before putting into transport Atanda et al. 609 containers.

Packaging

Use clean, smooth and ventilated containers for packaging. This is a very important factor in cutting down losses in these crops during harvesting, transportation marketing and storage. Use containers that are appropriate to the crop.

Transportation

Load and unload transport vehicles carefully. Use clean, well-ventilated vehicle covered at the top for transportation. Transport crops during the cool part of the day by driving carefully over smooth roads to minimize damage to crop. Attention must be given to commodity compatibility within a load. Ethylene-sensitive commodities such as lettuce should not be shipped with ethylene generators such as apples. A complete description of compatible and incompatible commodities is available (Ashby et al., 1987). The most common cause of freight claims is load shifting and crushing, but the costliest claims are the result of inadequate temperature control (Beilock, 1988). Other transportation steps are also important in quality maintenance, for example, from field to packing facility and from wholesale distribution point to retail outlet. The same principles that apply to longdistance ones, apply to short-distance shipments, but handling practices tend to receive less attention when the shipping distance is short. Fields and rural roads are usually bumpier than highways; thus, vehicles hauling the harvested crop from field to packaging house are generally not as capable of preventing shock and vibration damage as the tractor-trailer rigs will do. The delay of cooling of a crop is affected by the time required to load a vehicle in the field, the distance from field to packinghouse, the speed of the vehicle and the number of vehicles waiting to be unloaded at the packinghouse (Garner et al., 1987).

Storage

Only crops with high initial quality can be stored successfully, it is therefore essential to ensure that only crops of the highest quality (mature, undamaged) are stored. Separate damaged crops from wholesome crops and dispose of the damaged crop quickly several Nigerian fruits and vegetables cannot be stored in domestic refrigerator as they are susceptible to chilling injury. Examples are bananas, plantain and mango. Shelf life of a fruit or vegetable during storage is dependent on its initial quality, its storage stability, the external conditions and the handling methods. Shelf life can be extended by maintaining a commodity at its optimal temperature, relative humidity (RH) and environmental

conditions as well as by use of chemical preservatives or gamma irradiation treatment (Shewfelt, 1986). An extensive list of optimal storage temperatures and RI-Is with anticipated shelf life is available (Hardnburg et al; 1986). Controlled atmosphere storage is a conunercially effective means of extending the season of apples (Lidster, 1984). Atmosphere modification within wholesale or retail packages is a further extension of this technology. Modification of the atmosphere is being achieved by setting initial conditions and using absorbent compounds to limit carbon dioxide (CO2) and ethylene (C21-I4) concentrations (Kader et al., 1989; Labuza and Breene, 1989). They can however be stored in evaporative coolers as designed and constructed by Nigerian Stored Products Research Institute (NSPRI).

Waxing of the Surface

Waxing the surface of horticultural products is a treatment used on a number of commodities including citrus fruits, apples, rutabagas and cucumbers. It retards the rate of moisture loss and maintains turgor and plumpness and may modify the internal atmosphere of the commodity and is performed primarily for its cosmetic effect; the wax imparts a gloss to the skin and gives the produce a more shiny appearance than the unwaxed commodity. Sometimes, antiwaxing is a technique that could probably be used more widely in developing countries with advantage. In some countries, indigenous waxes may be suitable for this purpose. For example, experiments in Colombia have shown that waxing of cassava can extend the storage life from 2 to 3 days up to about 30 days by preventing discolouration in the vascular tissue. (Buckle et al., 1973) Work in India has also demonstrated the efficacy of indigenously produced wax emulsion formulations in extending the storage life of different fruits and vegetables (Dalal et al., 1970).

Environmental Effects

The waxes and wax formulations that are used in the U.S. are approved by the Food and Drug Administration and are kept under continuous review. Most of the ingredients in the wax mixtures are classified "generally recognized as safe" (GRAS). In most cases the skin is removed and discarded before consumption in which case the wax is not ingested and should cause no special problems. However, problems might arise if unregistered formulations are used or if the skin is eaten by humans or fed to animals.

Controlled Atmosphere Storage

Controlled atmosphere storage consists of placing a commodity is a gas-tight refrigerated chamber and allowing the natural respiration of the fruit to decrease the oxygen and increase the carbon dioxide content of the atmosphere in the chamber. Typically, for storage of apples the oxygen content is lowered to about 3% and carbon dioxide is allowed to increase to 1 to 5%. This atmosphere can extend the storage life of apples by several months and allows fresh apples to be marketed every month of the year. This technology requires expensive storage chambers and close supervision of the composition of the atmosphere and is unsuited for widespread use in less developed countries. Some roots and tubers are stored in pits in the ground, known as "clamp storage". Well-designed clamps tend to change the atmosphere to some extent by reducing oxygen and increasing the carbon dioxide content. Modified atmosphere storage would probably be effective for a limited number of commodities in developing countries especially if coupled with low temperature storage. Wills and Wimalasiri (Hort. Science, 14 528 1979) have recently shown that short pre-storage exposure to high carbon dioxide and low oxygen atmosphere of vegetables can extend the storage life of commodities even at ambient temperature.

Environmental Effects

Since this technology only manipulates the proportions of asses that are naturally present in the air there should be no adverse environmental effect. The now technology of hypobaric storage is emerging which maintains reduced pressure in the refrigerated storage chamber by means of vacuum pumps. In this system, the commodity is placed in a flowing stream of highly humidified air which is maintained at a reduced pressure and controlled temperature. Under these conditions, Bases released by the commodity that limits its storage life, are flushed away. Reports indicate that the storage life of certain fruits and vegetables is extended substantially by this procedure. The economic feasibility of this type of controlled atmosphere storage is presently being tested. This is an energy-intensive and capital-intensive technology and is perhaps unsuited for less developed countries. The major environmental effect is the high energy coat.

Field Factors

Maturity at time of harvest is an important factor in the keeping quality of horticultural products. Commodities that are harvested in an immature state not only have poor eating quality but may tend to shrivel in storage and be more susceptible to storage disorders. When picked too mature the commodity is soft or fibrous, the flesh breaks down more quickly and it has a shorter storage life. There is an optimum time of harvest to give maximum storage life for fruits, vegetables and tubers.

The rootstocks used for establishing fruit orchards may affect oases. For example, McDonald and Wutscher (1974) reported decay in grapefruit ranging from 3.3 to 27.7% depending on the rootstock. It is reported that the storage life of fresh cassava can be greatly extended by leaving part of the stalk attached to the tubers at harvest time. There are a number of other field factors that affect losses and these should be utilized as much as possible.

Environmental Effects

Generally, there are no adverse environmental effects in these operations.

Suberization and Curing

Potatoes, sweet potatoes, yams and several other roots and vegetables have the ability to heal skin wound when held at moderately warm conditions end high humidity for several days after harvest. The self-healing of wounds, cute and bruises is known as curing. There are two steps in the curing process. First is suberization; the production of suberin and its deposition in cell walls. The second is the formation of a cork cambium and production of cork tissue in the bruised area. The new cork tissue seals the cut or bruised areas and helps prevent the entrance of decay organisms. The healing of injuries received in harvesting and handling prolongs the storage time and reduces the incidence and spread of decay in storage.

The storage life of onions and garlic is extended by exposure to warm dry conditions for several days to dry the outside akin and prevent the ingress of spoilage organisms. This process if also known as curing although physiologically it is rather different and causes about 5% weight loss Curing is carried out in the field when weather conditions are suitable; otherwise the product is subjected to forced circulation of warm dry air when first put into storage.

This is sound environmental practice. There is little effect on the environment from curing.

Processing

Considerable quantities of fruit and vegetables are processed by dehydration, canning and freezing in developed countries. In developing countries, small amounts of these commodities are processed for local consumption, although large volumes of some commodities are processed for export (e.g., canned pineapple).

Canning and freezing require a high capital cost, high energy costs and expensive packaging and are unsuited for widespread use in less developed countries. Dehydration or sun drying is the simplest and lowest cost method of preservation and should be more widely used in developing countries because it converts a perishable commodity into a stable item with long storage life. Some excellent quality dehydrated products can be made from Atanda et al. 611 roots and tubers; this kind of processing should be encouraged.

Environmental Effects

Occupational hazards in the fruit and vegetable processing industry are the normal hazards associated with machinery, for which adequate safety measures are well developed. The national institute for occupational safety and health in the U.S. (NIOSH) has no complaints of safety hazards in processing plants that handle horticultural products. The U.S. Occupational safety and health administration (OSHA) have no regulations specific to the fruit and vegetable processing industry other than the board guidelines that apply to industry in general. The fruit and vegetable processing industry is not on the list of occupational groups in which excess cancer incidence is reported by the U.S. national cancer institute.

There are occupational risks to some workers with specific horticultural products. For example, Barber and Husting (1977) report isolated cases of contact dermatitis among workers handling raw fruits and vegetables, including carrots, asparagus, mangoes, cashew fruits and nuts and some citrus fruits. Fruit and vegetable handlers may also suffer contact dermatitis due to sensitivity to specific insecticides and fungicides. Indirect effects of handling fruits and vegetables include chapping and moniliasis from exposure to moisture, photosensitization dermatitis from sunlight and parasitism from mites. Products that cause photosensitization include fig, rue, lime, bergamot, paranips, parsley, carrots, fennel, dill and pink rot celery. Raw pineapple fruits contain the proteolytic enzyme bromelain which causes skin irritations to workers in pineapple processing plants. This problem is overcome by supplying workers who handle cut fruit with rubber gloves.

Fruits and vegetables can be preserved for relatively long period by drying, freezing and canning. Drying, freezing and canning are generally preceded by blanching for better retention of quality.

Sanitation

Strict hygiene is required at all stages of handling perishable crops. This will help to minimize infection by pathogens.

CONCLUSION AND RECOMMENDATIONS

It was revealed from the review that post-harvest food losses in perishable produce are product of many variables which are inter-woven. The factors contributing to these losses includes; the initial quality of the crop, mechanical injury, temperature, humidity, handling given to the crop and storage atmosphere. It was also observed that a substantial amount of post-harvest losses have their origin in the pre-harvest stage for example genetic factors, infections, pest-infestation, environmental factors and cultural practices during production stage.

In view of these factors, a good sanitation management in all pre and post-harvest operations in perishable crops will help in eliminating sources of infection and reducing levels of contamination. Proper timing and good methods of harvesting devoid of mechanical injuries will help in reducing food lose in perishable crops. It is clear that most post-harvest losses in perishable produce result from infection by fungi and bacteria (pre or post-harvest) and from inherent physiological activity although insect, rodents, nematodes and occasionally birds may cause significant losses under certain conditions.

- Post-harvest loss in perishable crops constitutes on important issue that needs increased and continuing attention at national regional and international levels by FAO, Government and other concerned organizations because it requires fewer resources and applies less pressure to the environment in maintaining the quantity and quality of food than through increase production to offset post-harvest losses.

- Department of post-harvest technology should be established in all faculties of agriculture in Nigerian Universities. This will ensure the availability of trained manpower to conduct effective post-harvest research.
- Good sanitation practices in all post-harvest operations is a very important factor in eliminating sources of infection and reducing levels of contamination.
- Research/extension activities to be geared up so that available technology on post-harvest food losses reduction is made known to all users.
- Traditional effective methods for preventing and reducing post-harvest losses need to be identified and exploited; this includes maintenance of continuous supply, storage for restricted periods and transformation to durable products. Some valuable traditional technologies for food preservation are in danger of becoming lost because they are being superseded by more sophisticated methods of doubtful long-term value. Modern and technology-intensive methods should be applied appropriately according to prevailing conditions including cultural factors. Efficient and proper management of such technologies is as important as the type of equipment and facilities selected.
- Most post-harvest losses in horticultural produce result from infection by fungi and bacteria (pre or post-harvest) and from inherent physiological activity although insects, rodents, nematodes and occasionally birds may cause significant losses under certain conditions, insects can disseminate some plant pathogens and also provide wounds as points of entry for micro-organisms. In general, pre-harvest application of fungicides in more important in the control of post-harvest problems of fruit and vegetable crops than in root crops.
- Techniques to reduce food losses require cultural and economic adaption. This is so because all food losses occur at a particular socio-cultural environment.

- There is a need to establish exchange programmes (technical co-operation between developing countries) between countries of similar needs and interest but which have apparent difference in advancement in post-harvest handling systems.
- The proper management of temperature and humidity of root crops and certain other perishable in the initial post-harvest period is essential to good curing which improved wound healing and minimizes infection by micro-organism.
- The use of post-harvest chemicals has not shown toxicological problems for now, but when they are used there is need to ensure that the dosages and residues conform to internationally recommended maximum levels e.g. of the FAO/WHO Codex Alimentarius Commission.
- Diagnostic studies using an interdisciplinary approach are needed to properly identify the area where losses occur within the post-harvest system of perishable food crops.
- The Nigerian stored products research institute (NSPRI) should be given sufficient funds to execute its research programme on effective crop storage in Nigeria.
- Scientists should come out with the production of a new packaging device capable of preserving food and cash crops such that the losses the nation has been witnessing in the agricultural sector will be eliminated.

REFERENCES

1. Ashby BH, Hinsch RT, Risse IA, Kindya WG, Craig WL, Turczyn MT (1987). Protecting Perishable Foods during Transport by Motor Truck. USDA Handbook No. 669. U.S. Govt. Printing Office, Washington, D.C.
2. Barber TE, Husting EL (1977). Plant and wood hazards. in Occupational Diseases, A Guide to Their Recognition. National Institute for Occupational Safety and Health, U.S. Department of Health, Education and Welfare. pp. 125, 126.
3. Beiioclt R (1988). Losses in the logistical system: The case of perishables. J. Food Dist. Res., II9(2): 20-28.

4. Buckle TS, Castelbanco H, Zapata LE, Bocanegra MF, Rodriguez LE, Rocha D (1973). Preservacion de yuca freasca for el metodo de parafinado. Instituto de Investigaciones Tecnológicas, Bogota, Colombia.
5. Burden OJ, Coursey DG (1997).Banana as a food crop. In:Leaky CLA, Wills JB (Eds) Food Crops of the Lowland Tropics London Oxford University Press, pp. 97-100.
6. Coursey DG (1976). The status of root crops: a culture historical perspective. J. Root Crops, 2(1): 1- 9.
7. Coursey DG, Booth RH (1977). Post-harvest problems of non-grain staples. Acta Hort., 53: 23-33.
8. Coursey DG (1978a). (In) www.fao.org/ x5045E05 html.
9. Coursey DG, Proctor FJ (1975). Towards the quantification of the postharvest [oases in horticultural produce. Acta Hort., 49: 55-63.
10. Coursey DG, Haynes PH (1970). Root crops and their potential as food in the tropics. World Crops, 22(4): 261-265.
11. Coursey DG (1983). Postharvest losses in perishable foods of the developing world. In "Postharvest Physiology and Crop Preservation" (M. Lieberman, ed.), Plenum Press, New York National Academy of Sciences (I978). "Postharvest Food Losses in Developing Countries." National Academy Press, Washington, D.C., pp. 485-514.
12. Dalal VB, Subrahmanyan H (1970). Refrigerated Storage of fresh fruits and vegetables. Climate Control, 3: 37.
13. Expert Consultation Meeting On Postharvest and Value Addition of Horticultural Produce (2010). GFAR/APARI Date: 29 November – 2 December 2010 http://www.apcoab.org/uploads/ExCon_PHTHort.pdf FAO (1979). Preliminary report on the Review Agriculture Towards 2000.
14. FAO (1981). Food loss prevention in perishable crops FAO AGRICULTURAL SERVICES BULLETIN No. 43.
15. Food and Agriculture Organization of the United Nations Rome 1981 www.fao.org/docrep/s8620e/S8620E07.htm.
16. Garner JC, Prussia SE, Shewfelt RL, Jordan JL (1987). Peach quality after delays in cooling. American Society of Agricultural Engineers Tech. Paper number B7-6501 St. Joseph, Michigan.

17. Hlardenburg RE, Watada AE, Wang CY (1986). The Commercial Storage of Fruits, Vegetables, and Florist and Nursery Stocks." USDA Agriculture Handbook No. 66. U.S. Govt. Printing Office, Washington, D.C.
18. Kader AA, Zagory D, Kerbel EL (1989). Modified atmosphere packaging of fruit and vegetables. Crit. Rev. Food Sci. Nutr., 28: 1-30.
19. Kuhr RJ (1979). Integrated peat management. A new strategy in an old war. New York's Food Life Sci., 12(2): 3-15.
20. Labuza TP, Breene WM (1989). Applications of "active packaging" for improvement of shelf-life and nutritional quality of fresh and extended shelf-life foods. J. Food Proc. Pres., 13: 1-69.
21. Lancaster PA, Coursey DG (1984). Traditional post-harvest technology of perishable tropical staples by Tropical Development and Research Institute (In) FAO AGRICULTURAL SERVICES BULLETIN No. 59 Food and Agriculture Organization of the United Nations Rome, 1984.
22. Lidster PD (1984). Storage of apples. J. Inst. Can. Sci. Technol. Alimenr., 17(3): 12-13. Atanda et al. 613
23. McDonald RE, Wutscher HK (1974). Root stocks affect poet-harvest decay of grapefruit. Hortscience, 9: 455-456.
24. National Academy of Sciences (1978). Post-harvest Food Losses in Developing Countries. National Academy of Sciences, Washington, D.C.
25. National Academy of Sciences (1978). Post-harvest Food Losses in Developing Countries. A Bibliography. National Academy of Sciences, Washington, D.C.
26. Nyamulinda B, Bizoza A, Rukazambuga D, Wanjiku C, Buruchara R, Mugabo J, Murorunkwere F Ntizo S, Musana BS, Ngaboyisonga C, Gafaranga J, Habumugisha P, Tuyisenge J, Birachi E, Adekunle AA, Fatunbi AO, Tenywa M (2011). Agricultural post-harvest Innovative technologies and access to niche market: Experience from Gataraga IP, Rwanda Learning Publics J. Agric. Environ. Stud., 2(1): 1-23.
27. Shewfelt RL, Prussia SE, Jordan JL, Hurst WC, Resurreccion AVA (1986). A systems analysis of postharvest handling of fresh snap hearts. HortSci., 21: 470-472.

28. Wills A, Wimalasiri W (Hort. Science, 14 528 1979). (In) Food loss prevention in perishable crops FAO AGRICULTURAL SERVICES BULLETIN No. 43.

Chapter 7

Changes in Physicochemical Attributes of Sweet Pepper (Capsicum annum L.) During Fruit Growth and Development

M.A. Rahman[1], G. M. A. Halim[2], M. G. F. Chowdhury[3], M.A. Hossain[4], and M. M. Rahman[5]

[1, 4, 5]Senior Scientific Officer

[2]Principal Scientific Officer, Horticulture Research Centre, Bangladesh Agricultural Research Institute (BARI)

[3]Scientific Officer, Postharvest Technology Division, B ARI, Gazipur, Bangladesh

ABSTRACT

Changes in physicochemical attributes of sweet pepper var. 'BARI Misti Morich-1' during fruit growth and development were studied. Capsicum flowers were tagged on the day of anthesis and fruits were harvested at three days interval started on 33 days after anthesis (DAA) until 48 DAA, when turned to light green colour. Length, diameter, and weight of fruits were linearly increased from 33 DAA to 48 DAA. Dry matter content, TSS, firmness, and shelf life of fruits were also gradually increased with lengthen the harvesting time from flower anthesis and reached from 5.1 to 7.3%, 4 to 5%, 7.1 to 14.5 kgf cm^{-2} and 9.6 days to 21.3 days, respectively, from 33 DAA to 48 DAA. However, ascorbic acid content of fruits was sharply decreased from 68 to 56 mg/100g from first harvest to last harvest, respectively. Hue angle declined with time while chroma and lightness values increased with fruit maturity. Considering all physical and nutritional characteristics and shelf life along with sensory evaluation scores, fruits of BARI Misti Morich-1 was found suitable to harvest after 45 days of anthesis, when they attained length 12.2 cm, diameter 8.8 cm, and weight 191g. Moreover, fruits were crispier, glossy with attractive colour and flavour and contained 58 mg/100g ascorbic acid at this maturity stage.

INTRODUCTION

Sweet pepper (*Capsicum annum* L.) is one of the most important high value crop grown extensively throughout the world, especially in the temperate region. The crop is highly sensitive to environmental factors like temperature and solar radiation (Bhatt *et al.*, 1999). However, efforts have been made for growing sweet pepper in the tropical and subtropical countries since last few years. In Bangladesh, some advanced farmers sporadically started to grow this crop to fulfill the local demand of urban areas (Saha and Salam, 2004).

Maturity is an integral component of produce quality particularly in the context of commercial maturity (Will *et al.*, 1998). According to Bautista (1990) 'maturity index is the sign or indication of the readiness of fruits and vegetables for harvest according to the consumer's choice'. Generally sweet pepper fruits are harvested at either the green mature or colouring stage. Usually field grown peppers are commercially harvested at the mature green stage (Lin *et al.*, 1993). Precise determination of sweet pepper fruit maturity is difficult at harvest. Fruits are harvested at different degrees of maturity and a proportion of fruits may be immature. Thus, harvesting of pepper fruits of different maturity at the same time is a common problem even though fruits may have the same skin colour at harvest (Tadesse *et al.* 2002). If fruits are picked immature, they may stay green longer but may not develop acceptable colour and flavour upon ripening (Boonyakiat *et al.*, 1987), which may lead to loss of consumer confidence. Moreover, since fruit growth continues until harvest, size of immature fruit will be smaller than more mature fruit resulting in a loss of harvest yield. Over matured fruit, on the other hand, lose their attractiveness and crispiness and become slimy in texture within very short time (Pareek, 2001). Therefore, determining the optimum maturity will benefit both consumer and grower. Normally, capsicum fruit maturity is assessed on the basis of subjective evaluations of visual colour (Lin *et al.*, 1993; Sanchez *et al.*, 1993; Howard *et al.*, 1994), size, and hand press to evaluate firmness or days from anthesis (Manandhar *et al.*, 1995; Sanchez *et al.*, 1993). Thus, there is a need to develop optimum maturity and quality indices for sweet pepper that are objective and consistent more reliable than existing methods.

A little information is available on the physiology of fruit growth and maturity of sweet pepper (Serrano *et al.*, 1995) and the changes of physicochemical attributes during maturation and ripening (Tadesse *et al.*, 2002). However, significant physiological changes associated with maturation and ripening may occur before or after commercial harvest and different cultivars may behave differently.

For determining appropriate harvest time, different attributes like surface colour, fruit firmness, soluble solids content, acid content, and concentration of volatile compounds as well as changes in other chemical constituents are used as indicators (Tadesse *et al.* 2002) so that products develop acceptable taste and flavour characteristics as well as maintaining structural integrity during subsequent storage and/or shelf life (Meredith *et al.*, 1989; Wills *et al.*, 1998). The present experiment was undertaken to study the changes in physicochemical attributes of sweet pepper during fruit growth and development and to determine a reliable harvest maturity index for the fruit.

MATERIALS AND METHOD

Crop Growing and Cultural Management

Sweet pepper fruits were obtained from field grown plants at experimental field of Horticulture Research Centre, BARI, Gazipur, Bangladesh. Seeds of sweet pepper var. 'BARI Misti Morich-1' were sown on 15 November 2010 in the seed bed. Seedlings with 2-3 leaf stage were transplanted into small plastic pots. Thirty five days old seedlings (4-5 true leaf stage) were then transplanted in the experimental plots. The experiment was carried out in a Randomized Complete Block Design with three replications. The unit plot size was 1m x 4m. Seedlings were placed in two rows of 50 cm apart with 50 cm between plants within a row to give a population of 16 plants/plot. Manure and chemical fertilizers were applied in the experimental plot as per recommendation (Saha and Salam 2004). The recommended dosage for one hectare area were 10 t cow dung, 100 kg N, 150 kg P_2O_5, 120 K_2O, 20 kg S, and 04 kg Zn. Half the quantity of cowdung was applied during final land preparation. The remaining cow dung, the entire quantity of P_2O_5, Zn, S, and 1/3rd each of N and K_2O were applied in bed during pit preparation. The rest of N and K_2O were top dressed in two equal splits after 25 and 50 days of transplantation. The intercultural operations, such as weeding, irrigation, and plant protection measures were taken as and when necessary.

Flower Tagging and Fruit Harvest

Flowers were tagged at anthesis to determine the stage of fruit development and 20 randomly selected fruits from 20 plants were harvested at three days intervals started on 33 days after anthesis (DAA) until 48 DAA to determine their growth and maturity. At each harvesting date, out of 20 fruits, 10 were used to determine physicochemical properties like fruit length, diameter, weight, firmness, crispiness, fibrousness, external colour, taste, and flavour, dry matter content, TSS, ascorbic acid content, and pH. The rest 10 fruits were packed in a 1% perforated polyethylene bag and held at 12 °C and 88 ± 5% RH in a constant temperature and humidity chamber (Model: VS-9111H-350, Korea) to observe the shelf life.

Measurement of Fruit Weight, Length, Diameter and Dry Matter

Fruit weight was determined using a digital electronic balance (Mettler PJ400, Switzerland) after transporting the harvested fruit in a closed plastic bag to the laboratory. Fruit length and diameter were measured with digital vernire caliper. For quantifying the dry matter content, the samples were kept in an electric oven at 70 °C for about 48h until a constant weight was achieved. Dry matter content was expressed as percentage which calculated based on its initial fresh weight.

Measurements of Surface Colour and Flesh Firmness

Surface colour of capsicum fruits was evaluated with a Chroma Meter (Model CR-400, Minolta Corp., and Japan). CIE $L^*a^*b^*$ coordinates were recorded using D65 illuminants and a 10° Standard Observer as a reference system. L^* is lightness, a^* (-reenness to +redness) and b^* (-blueness to +yellowness) are the chromaticity coordinates. The a^* and b^* values were converted to chroma $[C = (a^{*2} + b^{*2})^{1/2}]$ and hue angle $[h = \tan^{-1}(b^*/a^*)]$. Before measurement, the equipment was calibrated against a standard white tile. Skin colour was measured at three positions on the fruit surface, on the shoulder, at the equator and at the base. Three readings from each fruit were averaged prior to data analysis.

Flesh firmness of capsicum fruits was measured using a Digital Firmness Tester (DFT 14, Agro Technologie, France) fitted with a 5 mm diameter stainless probe at the locular space and carpel wall of the whole fruit at the equatorial region (Tadesse *et al.* 2002). Firmness was reported in kilogram-force/cm^2 (kgfcm^{-2}).

Measurements of Ascorbic Acid, Total Soluble Solids (TSS) and pH

Ascorbic acid content was estimated by 2,6-dichlorophenolindophenol titration following the method by Ranganna (1986), and results were expressed as mg100g^{-1} fresh weight. The total soluble solids and pH of the juice were determined by using hand-held Kruss refractometer (Model HR 900, SN 1200793, brix range0-90% at 20° C, Germany) and glass electrode pH meter (Delta 320, Mettler, Shanghai), respectively.

Evaluation of Sensory Attributes

Sensory evaluation, based on general visual appeals, colour, crispiness, flavour, taste, and visible structural integrity was conducted using a 7-point hedonic scale (Hernandez-Munoz *et al.*, 2008). The scores were: 7 = like extremely; 6 = like very much; 5 = like moderately; 4 = neither like nor dislike; 3 = dislike moderately; 2 = dislike very much and 1 = dislike extremely. Fruit scored above 4 was considered acceptable. Sensory evaluation was performed by a panel of judges consisting of 10 scientific personnel including both male and female members. Different preferences as indicated by scores were evaluated by statistical methods.

Statistical Analysis

The data were subjected to analysis of variance (ANOVA) using the CropStat Statistical Software version 7.2. The results showing significant differences were then subjected to mean separation using Tukey's Studentized Range (HSD) Test at $P \leq 0.05$. Analyzed data were graphically plotted using Sigma-plot 8.2 software.

RESULTS AND DISCUSSION

Fruit Length, Diameter, and Weight

Both length and diameter of capsicum fruits significantly ($P \leq 0.05$) increased with the increase of harvesting time from anthesis (Fig. 1). The longest fruit (12.28 cm) was harvested after 48 days of anthesis, which was significantly similar ($P \leq 0.05$) with fruits that harvested on 42 and 45 DAA. Whereas, the shortest fruits (9.78 cm) were obtained from harvesting after 33 days of anthesis. Similarly, fruits with maximum diameter of 9.14 cm were harvested after 48 days of anthesis, which was statistically similar with the fruits harvested after 45 days of anthesis. In an experiment Tadesse *et al.* (2002) found that capsicum fruit (cv. Domino) had attained almost 75% of its final length after three weeks of anthesis, but continued to elongate at a slower rate until 10 weeks after anthesis.

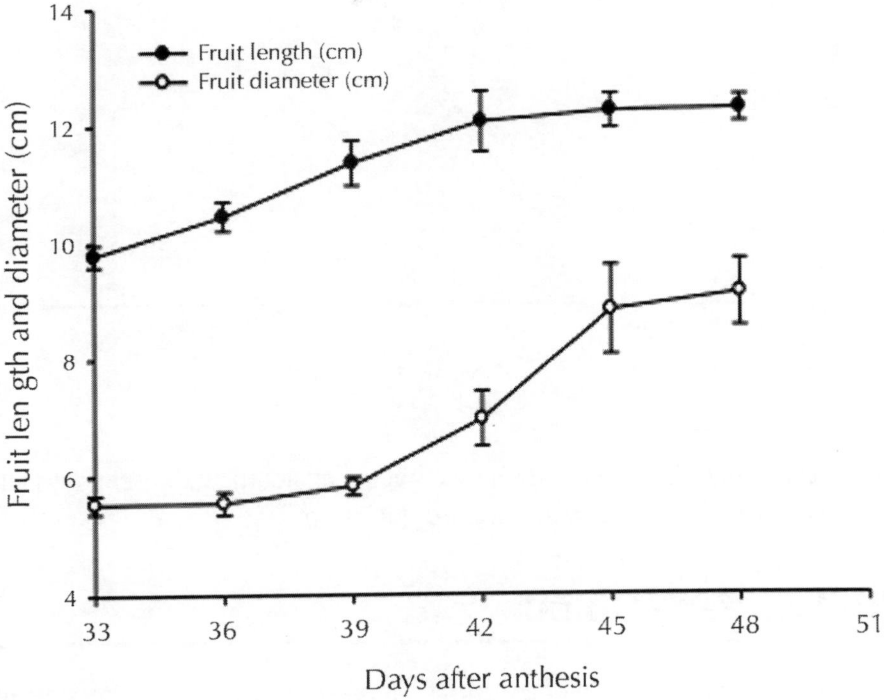

Figure 1: Changes in fruit length and diameter in capsicum at different days after anthesis. Vertical bars indicate standard deviation.

Capsicum fruits under this experiment showed gain in weight as extend the harvest time from flower anthesis (Fig. 2). The lighter fruits (78.26 g) were harvested after 33 days of anthesis, which increased steadily until 39 DAA. After this period, fruit weight was increased sharply with extend the harvesting time until 45 DAA (191 g). However, the maximum fruit weight of 192 g was obtained after 48 days of anthesis, which was statistically similar ($P \leq 0.05$) with fruit weight gained after 45 days of anthesis. Tadesse *et al.* (2002) reported that sweet pepper (cv. Domino) fruit fresh weight was linearly increased until eight weeks after anthesis.

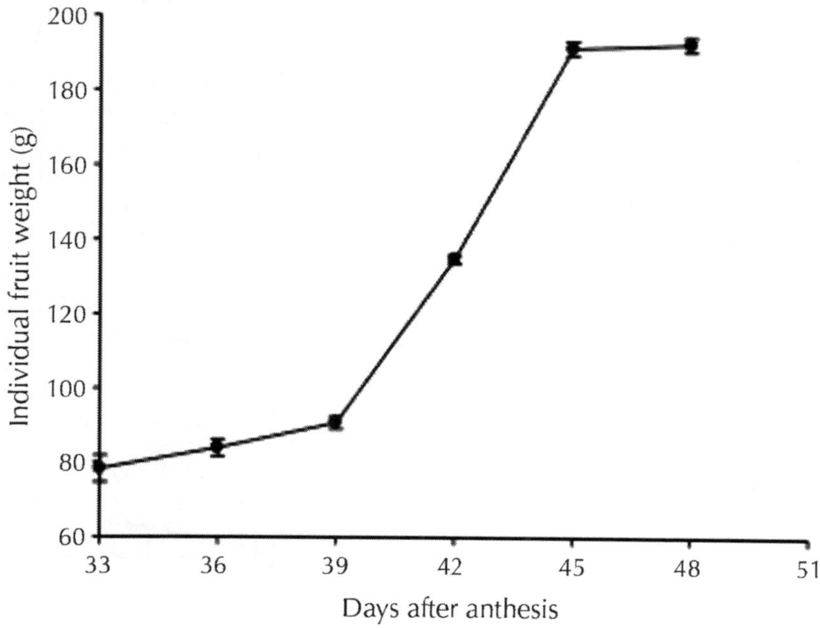

Figure 2: Changes in individual fruit weight in capsicum at different days after anthesis. Vertical bars indicate standard deviation.

Fruit Surface Colour

Surface colour changes of capsicum fruits were monitored by measuring lightness (L*), chroma (C*), and hue angle (h°) during harvesting periods. Values are presented in Fig. 3a-b, respectively. The intensity of green colour of fruit skin was slowly decreased with extend the harvesting time and turned to lighter green as evidenced by increasing values of L* and C* (Fig. 3a). The initial values of L* and C* were 35.12 and 21.52, respectively, were recorded in fruits that harvested after 33 days of anthesis which remained rather unchanged until 39 DAA. After that time, both were gradually increased and reached to 39.46 and 24.49, respectively, after 48 days of anthesis. There was no significant difference regarding the changes in L* and C* values among fruits harvested after 42, 45, and 48 days of anthesis. Results of the present study are partially supported by the findings of Tadesse *et al.* (2002) who observed a little increased in lightness and chroma of capsicum fruit during growth and development.

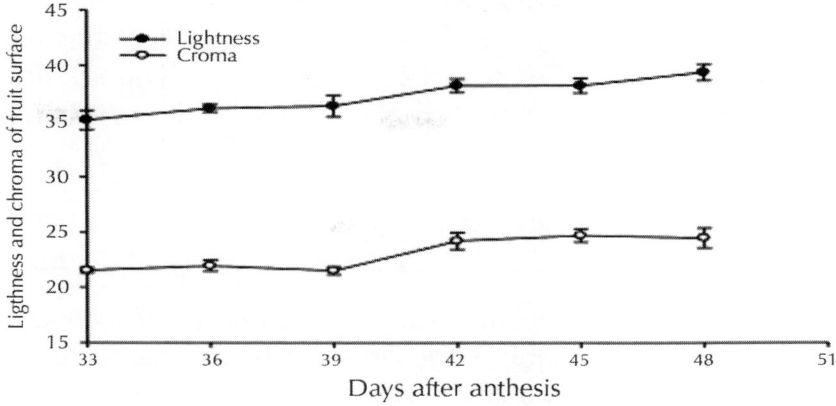

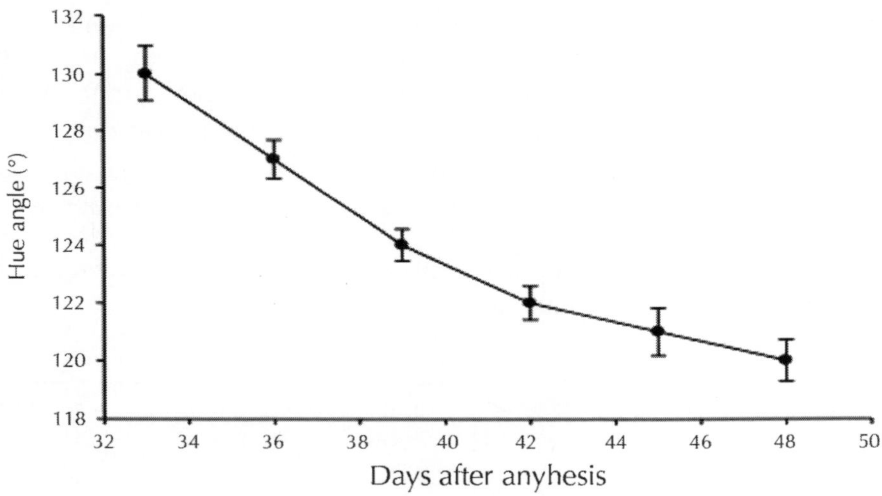

Figure 3: Changes in external colour (a) lightness and chroma; (b) hue angle of capsicum fruit at different days after anthesis. Vertical bars indicate standard deviation.

Hue angle is an indicator of colour change from green to yellow and red. In this study, the degree of hue angle was recorded 130 in fruit surface that harvested after 33 days of anthesis (Fig. 3b). After that, capsicum fruit showed a significant decrease in hue angle with extent the harvest period and attained the minimum level of 120 after 48 days of anthesis. This fall in hue value indicated that the intensity of green colour slowly decreased with maturity and turned to lighter green. Tadesse *et al.* (2002) found that the hue angle of capsicum fruits (cv. Domono) decreased sharply after seven weeks of anthesis that reflecting the colour change from green to red, which corroborates the present findings. The colour changes in sweet pepper correspond to a decrease in chlorophyll and an increase in carotenoid synthesis (Pretel *et al.*, 1995).

Flesh Firmness

Flesh firmness is a critical quality attribute in the consumer acceptability of fresh fruits and vegetables. The firmness of capsicum flesh that harvested after 33 days of anthesis was recorded 7.14 kgf cm^{-2} (Table 1), which progressively increased with extend the harvesting time from anthesis and reached the maximum firmness of 14.5 kgf cm^{-2} after 48 days of anthesis. Significantly similar ($P \leq 0.05$) firmness was found in fruits that harvested after 45 and 42 DAA. Result of this study supported the findings of Tadesse *et al.* (2002), who reported that pericarp thickness of capsicum increased with fruit age and was positively correlated with fruit firmness. The relationship between pericarp thickness and fruit firmness was also confirmed by Gu *et al.* (1999) in Jalapeno pepper.

Table 1: Changes in different physicochemical characters of capsicum fruits after different days of anthesis

Days after anthesis	Fruit firmness (Kgf cm-2)	Ascorbic acid (mg/ 100g)	Total soluble solids (%)	pH	Fruit dry matter (%)	Storage life (Days)
33	7.14d	68.0a	4.0d	5.79	5.19c	9.7d
36	10.0c	63.5b	4.2c	5.75	5.42c	15.7c
39	11.1c	61.0b	4.4c	5.73	6.14b	18.6b
42	12.3b	58.5c	4.7b	5.72	6.62b	20.3a
45	12.7b	58.0c	4.9b	5.72	7.21ab	21.2a
48	14.5a	56.0d	5.0a	5.71	7.35a	21.3a
F-value	*	*	*	ns	*	*
CV (%)	4.9	2.6	1.4	0.9	4.0	5.7

Values in each column followed by the same letter (s) are not significantly different at 5% level.

*Significant at 5% level; ns= Not significant

Ascorbic Acid, TSS, and pH

It is evident that the length of harvesting time induced significant variation in ascorbic acid content of capsicum fruits (Table 1). The ascorbic acid content was significantly higher (68 mg/100g) in younger fruits that harvested after 33 days of anthesis. But with the extend of harvest length, the content gradually decreased and reached the minimum level of 56 mg/100g after 48 days of anthesis. There was no statistical difference ($P \leq 0.05$) with regard to ascorbic acid contents in fruits harvested on 42 and 45 DAA. The losses of ascorbic acid can be continued through maturity, postharvest handling, processing, cooking, and storage of fruits and vegetables (Lee and Kader, 2000). Results of the present study are in agreement with Nagy (1980), who reported that immature fruits contained the highest concentration of ascorbic acid, whereas mature fruits contained the least.

The changes in TSS content of capsicum fruits with harvesting length is presented in Table 1. The TSS of fruit samples harvested after 33 days of anthesis was fairly low (4%), which was slowly increased with increase of harvesting time. The TSS content reached the maximum level of 5%, when fruits were harvested after 48 days of anthesis. No significant ($P \leq 0.05$) difference was observed regarding the TSS content in fruits that harvested between 45 and 48 DAA. The increase in TSS of sweet pepper fruit with maturity is probably a result of increased hexose sugar accumulation during fruit ripening (Nielsen *et al.*,1991) as there is a close positive correlation between the rise in TSS and soluble sugars (Mendlinger *et al.*, 1992). Results of the present experiment are in agreement with Tadesse *et al.* (2002) who found that the TSS of capsicum increased slightly from 5 to 6% during the period 2 to 6 week after anthesis.

The changes in pH of capsicum fruits as a function of different harvesting time are shown in Table 1. From the results, it was found that pulp pH of capsicum fruit was not affected by different harvest maturity. The pH ranged 5.71 to 5.79 was found during harvesting from 33 to 48 DAA, which was not significantly different among harvesting times. Similar results were also found by Fox *et al.* (2005) who reported that capsicum pH was not affected by different harvesting time, which ranged 4.9 to 5.1.

Dry Matter Content and Shelf Life

Dry matter content of capsicum was found lower (5.19 %) in fruits that harvested after 33 days of anthesis and then steadily increased as extend the harvest duration (Table 1). The maximum dry matter (7.35 %) was gained by fruits that harvested after 48 days of anthesis. Significantly ($P \leq 0.05$) similar dry matter was found in fruits that harvested on 45 DAA.

As changes in the dry matter content over harvesting period, similar trend was also observed for shelf life of fruits. The shelf life of 33 day's harvested fruits was only 9.7 days at 12 °C and 88 ± 5% RH, which steadily increased with increase the harvest duration and reached the maximum shelf period of 21.3 days for fruits that harvested after 48 days of anthesis (Table 1). However, no significant ($P \leq 0.05$) differences were observed in regarding the shelf life among fruits harvested after 45 and 48 days of anthesis. In this study, it was found that younger fruits which harvested after 33-36 days of anthesis had poor shelf life compared to the mature fruits. Results of this study are in agreement with Wills *et al.* (2007) who stated that respiration rate per unit weight is highest for the immature fruit or vegetables and then steadily declines with age and product that respires very fast will have a short storage life and vice versa.

Sensory Evaluation

Sensory data for capsicum quality based on visual appearance and organoleptic taste are presented in Table 2. Out of seven, the highest score (6.8) for overall sensory attributes was achieved by the fruits that harvested after 45 days of anthesis, which was significantly ($P \leq 0.05$) similar with fruits harvested after 42 days (6.6). At this stage of maturity, fruits were crisper, attained maximum flavour and taste with attractive skin colour. Therefore, judges showed a significant preference for fruits that harvested after 45 days of anthesis which obtained mean acceptance scores between 'like very much' and 'like extremely' on the hedonic scale.

Table 2: Overall acceptability based on visual appearance and organoleptic taste of capsicum fruit harvested at different maturity stage

Days of maturity	Overall sensory score (1-7)
33	4.8 ± 0.45d
36	5.0 ± 0.00c
39	6.2 ± 0.45b
42	6.6 ± 0.45a
45	6.8 ± 0.55a
48	6.2 ± 0.45b

Means within column with different letter are significantly different ($P \leq 0.05$)

Maturity of sweet pepper fruit was determined by a combination of different attributes. In this study, TSS, ascorbic acid content, surface colour change, and firmness were found to be good indicators for sweet pepper fruit maturity. Based on the results of the present study, sweet pepper fruits of the var. 'BARI MistiMorich-1' reached physiological maturity at 45 DAA. At this stage of maturity, fruit attained 12.24 cm length, 8.84 cm diameter, and 191 g in weight. Moreover, fruit contained 58 mg/100g of ascorbic acid, which is a very important quality attribute of fresh capsicum. Besides, fruits were more crispy, glossy, pleasant flavoured, and also tasty at this stage.

REFERENCES

1. Bautista, O. K. 1990. Postharvest Technology for Southeast Asian Perishable Crops. A Simplified Guide. Technology and Livelihood Resource Centre, UPLB, College Laguna, Philippines.
2. Bhatt, R. M., R. N. K. Srubuvasa and N. Anand. 1999. Response of bell pepper to irradiance photosynthesis, reproductive attributes and yield. *Indian J. Hort.* 56 (1):62-66.
3. Boonyakiat, D., P. M. Chen, R. A. Spotts and D. G. Richardson. 1987. Effect of harvest maturity on decay and postharvest life of `d'Anjou' pear. *Scientia Hortic.* 31: 131-139.
4. Fox, A. J., D. D. Pozo-Insfran, J. H. Lee, S. A. Sargent and S.T. Talcott. 2005. Ripening- induced chemical and antioxidant

changes in bell peppers as affected by harvest maturity and postharvest ethylene exposure. *Hort Science* 40(3): 732-736.

5. Gu, Y. S., L. R. Howard and A. B. Wagner. 1999. Physicochemical factors affecting firmness of pasturized Jalapeno pepper rings. *J. Food Quality* 22: 619-629.
6. Hernandez-Munoz, P., E. Almenar, V. D. Valle, D. Velez and R. Gavara. 2008. Effect of chitosan coating combined with postharvest calcium treatment on strawberry (*Fragaria x ananassa*) quality during refrigerated storage. *Food Chem.* 110: 428-435.
7. Howard, L. R., R. T. Smith, A. B. Wagner, B. Villalon and E.E. Burns. 1994. Provitamin A and ascorbic acid content of fresh pepper cultivars (*Capsicum annuum*) and processed Jalapenos. *J. Food Sci.,* 59: 362-365.
8. Lee, S. K. and A. A. Kader. 2000. Preharvest and postharvest factors influencing vitamin C content of horticulture crops. *Postharvest Biol. Technol.* 20: 207-220.
9. Lin, W. C., J. W. Hall and M. E. Saltveit Jr. 1993. Ripening stage affects the chilling sensitivity of greenhouse grown peppers. *J. Am. Soc. Hort. Sci.* 118: 791-795.
10. Manandhar, J. B., G. L. Hartman and T. C. Wang. 1995. Anthracnose development on pepper fruits with *Colletotrichum gloeosporiodes*. *Plant Dis.* 79: 380-383.
11. Mendlinger, S., A. Benzioni, S. Huyskens and M. Ventura. 1992. Fruit development and postharvest physiology of *Cucumis metuliferus* Mey, a new crop plant. *J. Hort. Sci.*67: 489-493.
12. Meredith, F. I., J. A. Robertson and R. J. Horvat. 1989. Changes in physical and chemical parameters associated with quality and postharvest ripening of harvested peaches. *J. Agric. Food Chem.* 37: 1210-1214.
13. Nagy, S. 1980. Vitamin C contents of citrus fruit and their products: a review. *J. Agric. Food Chem.* 28: 8-18.
14. Pareek, O. P. 2001. *Ber*. International Centre for Under-utilized Crops, Southampton, SO17 IBJ, UK.
15. Pretel, M. T., M. Serrano, A. Amoros, F. Riquelme and F. Romojaro. 1995. Non- involvement of ACC and ACC oxidase activity on pepper fruit ripening. *Postharvest Biol. Technol.* 5: 295-302.

16. Ranganna, S. 1986. Handbook of Analysis and Quality Control for Fruit and Vegetable Products. Second ed. Tata McGraw-Hill, New Delhi, India. Pp. 1112.

Chapter 8

Review on Postharvest Technology of Banana Fruit

M. Hailu[1], T. S. Workneh[2], and D. Belew[1]

[1]Harmaya University, College of Agriculture and Environmental Sciences, Dire Dawa, Ethiopia

[2]School of Bioresources Engineering and Environmental Hydrology, Faculty of Engineering, University of Kwa-Zulu Natal, Scottsville, Pietermaritzburg, 3209, South Africa

ABSTRACT

The aim of this review is in threefold: First, to explore the effect of different preharvest treatments on postharvest quality of fruits and vegetables. Second, the principles of biological, chemical and biochemical changes in banana during development, maturation, ripening and storage were reviewed. Third, postharvest handling and factors affecting quality of banana were examined. These include disinfecting, packaging and storage temperature. Pre- and postharvest treatments were found to have an effect on postharvest quality of banana, suggesting that postharvest quality of produce subjected to preharvest treatments should be assessed from a quality improvement, maintenance and consumer safety point of view. Literature recommends an integrated agro-technology approach towards improving quality at harvest and maintenance of qualities of banana.

INTRODUCTION

Banana, cooking banana and plantain (Musa spp. AAA, AAB and ABB groups) are major starch staple crops of considerable importance in the developing world. They are consumed both as an energy yielding food and as a dessert (Dadzie and Orchard, 1997). They are giant perennial herbs that originated in Southeast Asia. They evolved by natural hybridization between the two species M. acuminata (contributing genome A) and M. balbisiana (contributing genome B). All important bananas and plantains are triploid. Banana and plantain are monocotyledonnous plants belonging to the section Eumusa within the genus Musa of the family Musaceae in the order Scitamineae. Most edible-fruited bananas, usually seedless, belong to the species Musa acuminata Colla. Musa balbisiana Colla of southern Asia and the East Indies bears a seedy fruit but the plant is valued for its diseaseresistance and therefore, plays an important role as a "parent" in the breeding of edible bananas (Morton, 1987). Bananas and plantains are today grown in every humid tropical region and constitute the 4th largest food crop of the world after rice, wheat and maize (Picq et al., 1998; Arias et al., 2003). Moreover, with increasing urbanization, bananas and plantains are becoming more and more important as cash crops, in some cases providing the sole source of income to rural populations. Thus, playing an important role in poverty alleviation. Bananas and plantains are one of the cheapest foods to produce. The cost of production of one kg of plantain is less than that for most other staples, including sweet potato, rice, maize and yam. Consequently, bananas and plantains can be a very cheap food to buy and are, hence, an important food for low-income families (Picq et al., 1998). Bananas and plantains also grow in a range of environments and produce fruit year-round. Thus, it provides energy during the "hungry-period" between crop harvests. They are particularly suited to intercropping systems and to mixed farming with livestock and they are also popular as a backyard crop with urban populations. When grown in perennial production systems, they maintain soil cover throughout the year and if their biomass is used for mulch, soil fertility and organic matter remain stable. In mixed farming systems, bananas are used as a ground shade and nurse-crop for a range of shade-loving crops including cocoa, coffee, black pepper and nutmeg (Picq et al., 1998).

Banana ranks third place in world fruit volume production after citrus fruit and grapes at 64.6 Mt (FAO, 2000), and second place in trade after citrus fruit, at 14.7 Mt (FAO, 1999). Banana fruit is grown in more than 100 countries, mainly in sub-tropical areas (Stover and Simmonds, 1987). The biggest markets for banana are North Ameri-ca and Europe, followed by Japan and Eastern Europe (Caussiol, 2001).

Ethiopia lies entirely in the tropics where vast areas are suitable for banana growing. Banana production in Ethiopia ranges from homestead to large commercial plantations. At present, bananas are the leading fruit crops produced in the country both in terms of area coverage (28,695 ha) and production (1,245,615.60 q year^{1}) (CSA, 2004) where the bulk is produced in traditional agricultural system. Due to its relatively little requirement of land preparation, care, maintenance, and a comparatively high yield per given area and time, bananas are well suited to traditional agricultural system. It is recently reported that (Seifu, 1999) bananas are replacing other fruit crops such as lime and mango and the most important food crops like maize and sweet potato in some parts of the southern areas of the country.

In Ethiopia, research work on fruit and vegetables was started in early 1970s at major research centers, and cultivar development was the major research activity (Awole et al., 2011; Tigist et al., 2011). From introduced and locally collected varieties, Dwarf Cavendish, Poyo, Giant Cavendish and Ducasse hybrid were recommended for production (Seifu, 1999). Over all, banana research in the past and even at present, mainly focused on varietal development, selecting of disease resistant variety, fertilizer trial, clump management practices as well as irrigation trials in different centers. Even though theses preharvest practices are compulsory they must be coupled with the postharvest management practices because postharvest loses are of major concerns in many developing countries including ours (Tadesse, 1991; Workneh et al., 2010).

Ethiopia has a wide range of climate and soil types that enable it to produce a variety of agricultural products (Haidar and Demisse, 1999; Workneh et al., 2010). In addition, there is a vast potential of the internal market for bananas, primarily in densely populated urban areas, such as Addis Ababa. In addition, there is potential for export, being located close to important markets, such as Saudi Arabia, Djibouti, Somalia, etc. (Workneh et al., 2011). Despite these facts, marketing of fresh produce including bananas is complicated by postharvest losses both in terms of quantity and quality between harvest and consumption. Both the fresh market and processing plants for fruits are located in towns away from farms and producers in rural areas. The quality of the fresh and processed fruit depends on the postharvest handling during harvesting, transportation, and storage, and should be monitored effectively to keep the best quality of fruit at harvest. Due to a rain-fed farming system, lack of storage facilities, limited access to transportation, and risk of high losses, growers in Ethiopia are often forced to dispose off their produce over a short period of time (Haidar and Demisse, 1999; Workneh et al., 2010) which causes an economic loss of horticultural crops in general and fruits in particular.

Worldwide, postharvest losses in fruits and vegetables range from 25 to 40% (Raja and Khokhar, 1993) or even greater (Iqbal, 1996) in developed countries and up to 50% in developing tropical countries (Anon, 1989). Postharvest losses of fruits and vegetables in Ethiopia are estimated to vary between 25 and 35% (Tadesse, 1991). This high loss is due to lack of packaging, storage facileties, and poor means of transportation (Kebede, 1991; Workneh et al., 2010).There is no proper means of postharvest handling of fruits and vegetables at the retail and wholesale levels, which results in poor quality of fruits and vegetables at the consumer level. Although the country is experiencing huge postharvest losses of fruits and vegetables, very little or no emphasis is given to postharvest handling of perishable produces (Tadesse, 1991).

For the fresh bananas to reach the consumer in the right condition, it must be marketed properly, bearing in mind the application of most suitable temperature and humidity as well as appropriate packaging and handling methods. Good handling during harvesting can minimize mechanical damage and reduce subsequent wastage due to microbial attack (Wills et al., 1998).

Low temperature handling and storage are the most important physical method of postharvest management (Johnson et al., 1997). Mostly, the traditional packaging method for banana is nested packaging in which dried banana leaf are used but the effectiveness of these packaging materials even has not yet been well investigated.

Plastic films have also been found to increase the shelf life of banana fruit. Modified atmosphere packaging (MAP) of fresh fruit refers to the technique of sealing actively respiring produce in polymeric film packages to modify the O_2 and CO_2 levels within the package atmosphere. It is often desirable to generate an atmosphere low in O_2 and/or high in CO_2 to influence the metabolism of the product being packaged or the activity of decaycausing organisms to increase storability and/or shelf life. In addition to atmosphere modification, MAP vastly improves moisture retention, which can have a greater influence on preserving quality.

Furthermore, packaging isolates the product from the external environment and helps to ensure conditions that if not sterile, at least reduce exposure to pathogens and contaminants there extends the shelf life of the produce (Mir and Beadry, 2004).

So in Ethiopia, the research work on banana has focused mainly on varietal development and field management practices and little work has been done on postharvest management banana.

As a result, the quality of fruits is reduced and considerable amount is wasted, from harvesting to final consumption. This loss can be kept mini-mum by improving postharvest handling techniques. Therefore, this study aimed at reviewing the integrated agro technology of banana under Africa context fruit handling.

THE BANANA CROP

Banana production is restricted to tropical or near tropical regions, roughly the area between latitudes 30°N and 30°S (Morton, 1987). Within this band, there are varied climates with different lengths of dry season and different degrees and patterns of precipitation. A temperature range of 15 to 38°C occurs in most production areas, with the optimum temperature being 27°C (Nakasone and Paull, 1999). Yields are higher when temperatures are above 24°C for a considerable period. In cooler climate, the crop requires longer time to mature. Plants exposed to low temperature and humidity during active growth stage show reduced growth and yields. Banana crop requires a mean rainfall of 100 mm per month and there should not be more than 3 months of dry season (Morton, 1987).

Banana is just more than a crop. The ripe banana is utilized in a multitude of ways in the human diet from simply being peeled and eaten out of-hand to being sliced and served in fruit cups and salads, sandwiches, custards and gelatins; being mashed and incorporated into ice cream, bread, muffins, and cream pies. Ripe bananas are often sliced lengthwise, baked or boiled, and served (perhaps with a garnish of brown sugar or chopped peanuts) as an accompaniment for ham or other meats. Ripe bananas may be thinly sliced and cooked with lemon juice and sugar to make jam or sauce, stirring frequently during 20 or 30 min until the mixture jells. Whole, peeled bananas can be spiced by adding them to a mixture of vinegar, sugar, cloves and cinnamon which has boiled long enough to become thick, and then letting them cook for 2 min (Morton, 1987). On the other hand, banana is the cheapest source of carbohydrate. In addition, ripe banana fruit is a good source of vitamins (A, B and C) and minerals (Ca and Fe) and it is a known source of potassium (Samson, 1991).

PREHARVEST FACTORS THAT"INFLUENCE THE POSTHARVEST QUALITY OF BANANA

The quality of fresh fruits and vegetable offered to consumers is constrained by the level of quality achieved at harvest, and generally cannot be improved by postharvest handling, rather can be maintained. Myriads of preharvest genetic and environmental factors affect the growth, development and final quality of fresh fruits and vegetables (Shewfelt and Prusia, 1993).

Genetic Factors

Banana breeding has been existing for more than seventy years (Ortiz et al., 1995). Smith (1995) suggested that future developments in the banana fruit sector would depend upon cultivar selection, plant breeding and genetic engineering. Varieties differ in many characteristics, including visual appearance (example size), yield and quality. Size, for example small, medium or large, is a matter of consumer preference (Hofman and Smith, 1993). Variety also has an effect on yield, firmness, fibrousness, succulence and juiciness (Kader, 1992). For certain tree crops, rootstock selection may cause differences in fruit total soluble solid (TSS) and acidity via influences on nutrient and water uptake and translocation or differences in photosynthate partitioning (Beverly et al., 1992). Increasing the energy supply and decreasing the water content of fruit increases TSS in tomatoes. Thus, TSS exemplifies a tradeoff between yield and quality, since yield generally decreases with increasing TSS (Shewfelt and Prusia, 1993). The genotypic characteristics of any one cultivar vary in response to environmental effects.

Canopy Management

Canopy management focuses on the amounts of light and CO_2 that fruits receive. For banana fruit, full shade gives a dull yellow peel colour whereas partial shade leads to a bright yellow peel colour. Low light intensity retards development of carotenoids (Caussiol, 2001). An important determinant of banana fruit quality is row spacing and the associated plant population. Plant density consists of selecting the most vigorous suckers located in the best places and eliminating undesirable ones. This method can increase the number of leaf and fruits exposed to sunlight. Removal of leaf can also help prevent fruit scaring. Bunch thinning reduces inter-fruit competition and improves fruit size. An average banana plant population is around 2, 500 per ha (Stover and Simmonds, 1987). Plant health and leaf/fruit ratio also influences flavor (Hofman and Smith, 1993).

Climatic Condition

Temperature is the most important factor in affecting growth and development of banana crop. Bananas flourish under uniformly warm to hot conditions. Shoot growth is best between 26 to 28°C and fruit growth at 29 to 30°C. Plant growth slows below 16°C and stops at 10°C. The optimum for dry matter accumulation and fruit ripening is about 20°C and for the appearance of new leaf about 30°C (Nakasone and Paull, 1999).

Theoretical determination of the harvesting stage of Cavendish banana showed that increased fruit diameter in the absence of limiting soil factors is a function of the total daily temperature from the appearance of the last hand of the bunch up to the harvest stage (¾ full or 34 mm grade). The total daily temperature is defined from a 14°C threshold, which represents the minimum temperature permitting the growth of bananas. In optimal conditions without constraints other than temperature, a total cumulative temperature of 900°C is required to obtain the theoretical optimum harvest stage of dessert bananas (Chang et al., 2000).

Also, a relationship between natural production conditions and physical fruit traits (texture and colour) were reported by Bugaud et al. (2007) that the peel of green bananas harvested during the hot humid season was not as hard as that of bananas harvested during the cool dry season. In ripe bananas, a decreasing correlation was also noted between the mean daily temperature and the fruit yellowness. This interaction could be responsible for the yellower colour banana pulp and of bananas harvested during the coolest seasons. The green life of bananas harvested during the hot humid season was shorter than that of bananas harvested during the dry and intermediate seasons (Bugaud et al., 2007).

The effect of high temperature is the function of time in which at 48°C the damage occurred in 3 h and 45 min, at 57°C it took 7 min and at 62°C it took only 1 min (Gowen, 1995). Under peel discolouration, the fruit was observed with air temperature of 33 to 35°C but was associated with high light intensity and moderate to low rainfall. A more widely occurring effect of high temperature is bleaching of the chlorophyll in both leaf and fruits.

Another important effect of temperature is chilling damage during crop development. It includes reddish brown streaking on the skin of the fruit, reduced latex flow, slow colour development of the fruit during ripening, a dull or grayish yellow colour in ripe fruits and brown skin in advanced stages. More subtle changes are reduced production of volatiles during ripening and disturbances to the evolution of CO_2 and ethylene (Gowen, 1995). In addition, Crane et al. (2005) reported that a temperature below 16°C but above 0°C resulted in failure of the flowering stalk or fruit bunch to emerge from the pseudostem (called choking) and an increase in fruit rotting during ripening.

Water Management

Banana plant must be irrigated immediately after planting and about 30 to 40 weekly irrigation is required. Inadequate irrigation to banana leads to delayed flowering, bunch size, delayed maturity of reduced fingers, induces physiological disorders and also poor keeping quality of fruits (Lopez, 1998). However, excess water also has detrimental quality consequences for plant. The photosynthetic rate decreases with overly high water availability and low transpiration rates. High moisture content in fruit also tends to dilute the soluble solids leading to low flavor intensity. During dry periods, soluble solids accumulate rapidly inside plant structure. When the water supply increases suddenly, by irrigation or rainfall, water moves quickly to the plant cells in response to the osmotic gradient. The rapid rise in hydro-static pressure can rupture cell membranes and walls, leading to splitting of fruits and other structures (Shewfelt and Prussia, 1993). Furthermore, a high relative humidity during fruit development shortens the storage life and increases the incidence of finger drop and crown rotting (Munasque et al., 1990). Generally, the water regime in the early stage of growth had a more pronounced effect both on dry matter accumulation and nutrient uptake than it did after bunch emergence (Gowen, 1995).

Nutrient Management

For growth and fruit production, bananas require high amount of nutrients which are often supplied only by the soil. Since large amount of nutrients are removed from the soil and these nutrients have to be replaced to permit continuous production of high yield and this is achieved through application of organic manure and/or mineral fertilizers (Gowen, 1995).

Nitrogen is considered second only to potassium for growth and production. Beyond its effect on growth and production, it also affects bunch maturation period and fruit quality in which as nitrogen level increases the length of the maturation period also increases. In addition, nitrogen also affects bunch weight through its effect on number of hands and fingers, fruit circumference and weight. It also affects fruit quality by reducing total sugars and total soluble solids, and increasing acidity (Gowen, 1995).

On the other hand, potassium is found to be the key element in banana nutrition. It increases both the yield and quality of fruits. High potassium and calcium will give high dry matter and glucose content in the peel and pulp of banana fruit (Caussiol, 2001). Respiration was lower in potassium deficient plants and therefore the main effect of low potassium supply on dry matter production would be through reduction of photosynthesis. On the contrary, low levels of nitrogen, phosphorus and magnesium give high dry matter in the pulp. Low potassium produces thin fruits and fragile bunches. Excess potassium over nitrogen causes a premature ripening of fruits (yellow pulp). In addition, a potassium supply above that which influence growth and yield, changes in reducing and total sugars have reported. As potassium supply increases, the amount of sugar will increase and the acidity decreases. Thus potassium supply has an effect on fruit quality over and above its influence on yield (Gowen, 1995).

Phosphorus requirement of banana is not large as opposed to its demand for potassium and nitrogen, this can be explained by the facts that bananas accumulate the phosphorus they require over extended period of time; and a relatively small quantity of phosphorus is exported with the fruit and that phosphorus is easily redistributed from old to young leaf (Gowen, 1995).

Stage of Maturity and Ripeness

The bananas are harvested at a specific maturity stage in which the maturity indices are based on the age of the bunch, the interval between flowerings and harvesting, the filling of the fingers or the colour of the skin and pulp and brittleness of the flower end of the bunch. The filling of the fingers is the criterion mostly used. This standard is typically completed by other visual criteria like the evolution of the peel colour of the fruits. Most of these criteria depend on the cultivar. If the filling of the fingers was combined with the colour of the pulp, evaluating harvest maturity could become more objective (Chang et al., 2000). The interval between flowering and harvesting is also an objective criterion, which can be grounds for harvest decision.

In practice, harvest maturity of fruits will depend mostly on the target market or the duration of the required postharvest life. Banana to be marketed locally can be harvested at a more mature stage than those which are to be exported. They are harvested at 75 to 80% maturity for long distance transport taking some three weeks and at full mature stage for local domestic market (Gowen, 1995). If the fruits are to mature at harvest they may well split during handling, especially if they have recently been irrigated or it has rained.

According to the study of Ahmad et al. (2007), speed of ripening was independent of harvest maturity at the ripening temperature of 16°C but that it was reduced at 13°C for more immature fruit. Also chilling injury occurred at 13°C which reduced the acceptability of the fruit. The symptoms of chilling injury progressed as the fruit ripened and were more apparent at colour stage 6 than early colour stages. Bananas from the early stage of maturity showed increased levels of chilling injury symptoms. Early harvesting resulted poor quality ripe fruit in terms of the total soluble solids and eating quality. Therefore, the variation in the quality of ripe fruit on a commercial scale could be due to the difference of harvest maturity.

POSTHARVEST PHYSIOLOGY OF BANANA

Respiration

Respiration is a process by which stored organic materials (carbohydrates, proteins, and fats) are broken down into simple end products with a release of energy. Oxygen is used in this process and carbon dioxide is produced. Under normal atmospheric conditions, aerobic respiration takes place. During the respiration process there is a loss of stored food reserves in the commodity. This leads to hastening of senescence because the reserves that provide energy are exhausted. Also, there is loss of flavor quality, especially sweetness. The rate of deterioration of harvested commodities is generally, proportional to the respiration rate (Irtwange, 2006).

Bananas are harvested green and ripened in market areas. Fruits that are allowed to ripe on the tree often split, and tend to be mealy. The preclimacteric period after harvesting is vitally important for importers and ripeners because banana is transported before it is ripened. During this period, mature green fruit have a low basal respiration rate and ethylene production is almost undetectable. This period is also called the "green life" and the longest practical preclimacteric period is desired. Green life can be extended by decreasing temperature to 14°C, and storage under low O_2 (≤ 8%) and high CO_2 (≥ 2%) (Marriott and Lancaster, 1983).

Since banana is a climacteric fruit it exhibits a respiretory peak during ripening, after harvest at 20°C. Within a couple of days, respiration rate of about 20 mg $C0_2$ Kg^{-1} h^{-1} in the hard green banana fruit may rise to about five times at the climacteric peak and then fall as the ripening advances and there is also a considerable water loss through transpiration after the initiation of ripening (Salunkhe and Kadam, 1995).

The loss of substrate from stored plant product result in a decrease in energy reserves with the tissue, which in turn decrease the length of time the produce, can effectively maintain its existing condition. Respiration also reduces the total food. Finally, in marketing systems based on weight, respiratory losses of carbon represent weight losses in the product; hence, a decreased value (Kays, 1997).

Another extremely important effect of respiration is the removal of oxygen from the storage environment. If the ambient oxygen concentration allowed being excessively depleted, anaerobic conditions occur that rapidly spoil many plant products. As, a consequence, the rate of respiration is important for determining the amount of ventilation required in the storage area. It is also critical in determining the type and design of packaging material that can be used. The respiratory reduction in oxygen concentration in the storage environment can also be used as a tool to extend the storage life of a product. On the other hand, elevated ambient levels of carbon dioxide resulting from the respiratory process can be used to decrease respiration since its accumulation impedes the rate at which the process proceeds (Kays, 1997). The two major respiratory substrates found in fruit are sugars and organic acids.

Both sugars and organic acids are found largely sequestered within the vacuole, and form a major contribution to the overall flavor of the fruit and the balance between sugars and organic acid can markedly affect the taste of the fruit. Both sugars and organic acids originate from photosynthetic assimilates. Fruits differ in how they assimilate, accumulates during development and ripening. Some fruits have accumulated the bulk of their carbohydrate prior to the onset of ripening. This is either stored primarily as starch, as in banana, or as sugars, as in tomato. Fruits which accumulate their assimilate prior to ripening can be harvested at the mature green stage and still attain acceptable flavor on ripening. This is important when considering the need for early harvest to optimize shelf life (Tucker, 1993).

Effect of Ethylene on Ripening

Ethylene production rates increase with maturity at harvest, physical injuries, disease incidence, increased temperatures up to 30°C and water stress (Pesis, 2004). On the other hand, ethylene production rates by fresh horticultural commodities are reduced by storage at low temperature, by reduced O_2 levels, and elevated CO_2 levels around the commodity (Irtwange, 2006). Exposure of climacteric fruits to ethylene advanced the onset of an irreversible rise in respiration rate and rapid ripening. Various packages can delay the onset of climacteric and prolong shelf life of fruits by reducing ethylene production and sensitivity.

Ethylene appears to be intimately involved in the initiation of ripening in banana, as in other climacteric fruit, but its mode of action is unknown (Seymour, 1993). Unripe banana show a constant, but low level of ethylene production until the onset of ripening. Ethylene production then increased and is followed by a rise in the rate of respiration. During ripening peak ethylene is normally reached while the rate of respiration is still increasing. As the rate of ethylene production declines, the rate of respiration reaches its maximum at around 125 mg CO_2 Kg^{-1} and then declines slightly, but remains at high level (Seymour, 1993). So to achieve optimum fruit quality, postharvest techniques like the application of controlled atmospheric storage and modified atmospheric packaging are used in order to modulate the physiological processes of ripening of banana fruits.

Transpiration Loss

According to the study of Irtwange (2006) transpiration is the evaporation of water from plant tissues. Water loss is a very important cause of produce deterioration, with severe consequences. Water loss is, first, a loss of marketable weight and then adversely affects appearance (wilting and shriveling). Also, the textural quality is reduced by enhanced softening, loss of crispness and juiciness, and reduction in nutritional quality. The nature of the epidermal system of the commodity governs the regulation of water loss that is affected as well by environmental factors. Eventually, transpiration is a result of morphological and anatomical characteristics, surface-to volume ratio, surface injuries and maturity stage on the one hand, and relative humidity, air movement and atmospheric pressure on the other hand. As a physical process, it can be controlled by applying waxes and plastic films as barriers between the produce and the environment, as well as by manipulating relative humidity, temperature and air circulation.

The banana skin bears stomata and transpiration continues after the bunches has been cut. The magnitude of transpiration depends on temperature and relative humidity and absolute values for it are not worth reporting here for this reason. In trend, however, it shows a very marked relation to ripening which may briefly be described as follows. The green fruit, immediately after cutting, shows an initial fall in transpiration rate and then settles down to a steady state at level depending upon temperature and humidity; at the climacteric there is a sharp peak followed, as the fruit ripens, by the attainment of a new steady state (Simmonds, 1959).

There is usually a final rise in water loss which is related to degenerative changes of the skin caused by fungal attack; since the skin in this stage is senescent the loss can hardly be described as transpiration. These facts were established for individual fruits; because the bunches ripens progressively from the top hand to the bottom, the climacteric transpiration peaks of individual fingers are concealed and whole bunches (or bigger bulks of fruit) show simply a steady rising curve of water loss during ripening (Simmonds, 1959).

POSTHARVEST BIOCHEMISTRY OF BANANA

The ripening process, at the fruit level involves several biochemical pathways like degradation of starch to sugar, change in the peel and pulp colour, cell wall changes, change in the concentration of volatiles and acids (Gowen, 1995).

Texture

Most fruit soften during ripening and this is a major quality attribute that often dictates shelf life. Fruit softening could arise from one of the three mechanisms: Loss of turgor; degradation of starch; or breakdown of the fruit cell walls. Loss of turgor is largely a non-physiological process associated with the postharvest dehydration of the fruit, and as such can assume commercial importance during storage. Degradation of starch probably results in a pronounced textural change, especially in those fruit like banana, where starch accounts for a high percentage of the fresh weight (Tucker, 1993; Turner, 2001).

Carbohydrates

Starch forms about 20 to 25% of the fresh weight of the pulp of unripe bananas. During ripening this starch is degraded rapidly and the sugars sucrose, glucose and fructose accumulate; traces of maltose may also be present. Sugars are present in the green fruit only very small amounts, average about 1 to 2% of the fresh pulp; they increase to 15 to 20% at ripeness, the beginning of the increase coinciding with the respiration climacteric. Starch disappears concurrently, dropping from about 20 to 25% in the green fruit to about 1 to 2% in the ripe fruit; it is higher in the ripe plantain (6%) than in the dessert bananas (Simmonds, 1959). In the banana pulp sucrose is the predominant sugars, at least at the start of ripening, and its formation precedes the accumulation of glucose and fructose. The peel tissue also contains starch, about 3% fresh weight, and appears to show similar changes in carbohydrate during ripening. These characteristic patterns of carbohydrate metabolism can be altered under certain environmental conditions such as exposure to elevated temperatures during ripening (Seymour, 1993; Turner, 2001).

Pigments

There is a decrease in chlorophyll content from between 50 to 100 μg per gram fresh weight to almost zero in ripe fruit, while caroteniod levels (xanthophylls and carotene) remained approximately constant at 8 μg per gram fresh weight. There is reduction in total caroteniod content in the peel during the early stage of ripening followed by caroteniod biosynthesis at the yellow-green to yellow-ripe stage (Seymour et al., 1987).

Cell Wall Change

Softening of fruits appears to be closely linked with changes in their cell walls structures. In bananas, the changes in texture of the fruit during ripening probably result from alterations in both cell wall structure and the degradation of starch (Seymour, 1993).

Hemicellulose and pectin are components of cell wall and their concentration during ripening follows a trend similar to that of starch, dropping from 7 to 8% of the fresh pulp in the green fruit to about 1% at ripeness. There are indications of interconversion of starch and hemicellulose in the early phases of storage of green fruit. In behavior, however there is an important difference from starch, for the hemicellulose disappear whether or not the fruit is normally ripened; in fruit kept in prolonged cool storage, starch concentration declines slowly so that a high starch content is characteristic of "chilled" fruit until a very advanced stage of ripeness. The hemicelluloses, however, disappear at the normal rate under such treatment and their fate is still mysterious and they may be connected with loss of dry weight of the fruit which, it will be remembered, excess the loss that can be accounted for by respiration (Simmonds, 1959).

Organic Acids

Pulp pH and total titratable acidity are important postharvest quality attributes in the assessment of fruit ripening quality. Generally, when fruits are harvested at matured green stage, the pulp pH is high but as ripening progresses pH drops. Thus, the pulp pH could be used as an index of ripening (Dadzie and Orchard, 1997). The skin of the fruit shows similar trend but is slightly delayed with respect to the pulp; this is not surprising in view of the fact that ripening proceeds from the core of the pulp outwards. Various measurements of pH range between about 5 to 5.8 for the pulp of the green fruit and between about 4.2 and 4.8 for post climacteric fruit (Simmonds, 1959).

The levels of organic acids in a fruit can markedly affect its taste. Generally, in banana the pulp shows an increase in acidity during ripening and the main organic acid present are malic, citric and oxalic. As ripening advances, acidity declines presumably due to the utilization as respiratory substrates. The astringent taste of unripe bananas is probably attributed at least partly to their oxalic acid content, which undergoes significant decarboxylation during ripening probably by the action of oxalate oxidase (Seymour, 1993). Despite the increase in acid with ripening, is less that the rate of increase of sugars and total soluble solids (Gowen, 1995; Abbas et al., 2012).

Enzyme Activity

Banana fruit contains several hydrolytic and oxidative enzymes. The relative activities of α-mylase, starch phosphorylase, acid phosphate, and catalase increased considerably in banana fruits stored for 5 weeks at 20°C. The stage of maturity of banana fruit at harvest significantly influenced most of the physical and biochemical constituents and the activities of some enzymes. Banana fruits harvested at mature or early mature stage had a longer storage period and better quality. The potential storage life of banana fruits decreased with an advance in stage of maturity (Salunkhe and Kadam, 1995).

POSTHARVEST MICROBIOLOGY OF BANANA

Postharvest diseases can cause serious losses of fruits both in terms of quantity and quality. Fruits infected with disease have no market value. There are many postharvest diseases of banana, cooking banana and plantain. Among these important ones are crown rot, anthracnose and cigar-end rot (Dadzie and Orchard, 1997).

Crown Rot

Crown rot is one of the most important postharvest diseases of banana/ plantain. It is characteristically a disease complex caused by several fungi, sometimes in association with other micro-organisms such as bacteria. The most common pathogens associated with crown rot are Colletotrichium musae (Gloesporium musarum), Fusarium roseum, Fusarium semitectium and Botryodiplodia theobromae. When the hands are cut from the stems, the massive open wound is an ideal weak spot for crown rot fungi to enter and grow. Fungal spores on the fruit in the field are carried along (after harvesting bunch) to the packing house. Spores follow the fruit right into delatexing baths, where they are drawn deeply into the weak spot, the wound on the crown tissue (due to dehanding). Spores also remain on the outside of the fruit and are packed (Dadzie and Orchard, 1997; Dionisio, 2012). Symptoms of crown rot include softening and blackening of tissues at the cut crown surface, white, grey or pink mould may form on the surface of the cut crown and infected tissue turns black and the rot may advance into the finger stalk. The control of crown rot starts in the field with the regular removal of leaf trash. Proper field sanitation can greatly reduce the number of crown rot fungi spores present and not keeping rotting fruits or plant waste materials near the packing station. Dehanding should be done carefully with a sharp knife so as to avoid leaving a ragged cut. Finally, postharvest treatment of fruits with an effective fungicide is essential (Gowen, 1995; Dionisio, 2012).

Anthracnose

Anthracnose is one of the important postharvest diseases of banana, cooking banana and plantain. It is caused by the fungus, C. musae. Anthracnose is common on wounds, but it is capable of attacking sound fruit as well. Occasionally, it invades the necks of the fingers when they are damaged. It is characterized by numerous small circular and brown to dark brown spots (Gowen, 1995; Dadzie and Orchard, 1997). Preventive measures begin in the plantation. It is important to maintain strict hygiene or sanitation in the plantation and pack house, in order to minimize the number of spores available for infection. All cultural practices that reduce scarring and injury to the fruit will prevent anthracnose (Dadzie and Orchard, 1997).

Cigar-End Rot

Cigar-end rot of banana and plantain is an important postharvest disease caused by the fungi, Trachysphaera fructigena. Cigar-end rot is essentially a plantain disease (but it is also found in banana and cooking banana), the fruits being apparently most subject to attack in their more immature stages. The number of fingers affected in the bunch varies. The infection which starts with localized darkening and wrinkling of the skin originates in the perianths and spreads slowly backwards along the finger. The darkened area is bordered by a black band and a narrow chlorotic region which separates infected and healthy tissues. The principal method of control is frequent manual removal and burning of dead flower parts and infected fruits. Use of fungicide to control the disease is also recommended. In the pack house, care should be taken to cull infected fruits to avoid contaminating the washing water with spores. Cigar-end rot is effectively controlled by covering the flower (immediately after emergence) with a polyethylene bag before the hands emerge (Dadzie and Orchard, 1997).

POSTHARVEST HANDLING

According to the study of Kader (1992), as cited by Irtwange (2006), microbiological, mechanical and physiological factors cause most of the losses in perishable crops. Other causes of losses, according to the study of Fallik and Aharoni (2004) as cited by Irtwange (2006) may be related to: inadequate harvesting, packaging and handling skills, lack of adequate containers for the transport and handling of perishables, storage facilities inadequate to protect the food and transportation inadequate to move the food to market before it spoils.

Banana fruit is very susceptible to both abrasion and impact injury. The major marketing problem of export bananas is mechanical injury, which shows itself as black sunken areas on the skin after ripening. So during harvesting, care should be taken not to let the bunch fall on the ground and get damaged. During transportation the thin plastic liner in export cartons minimize chafing damage to fingers that rub against the side of the carton during handling. Latex allowed to dry on the skin oxidizes as brown and black stains and can also lead to downgrading of fruit. Other concerns are diseases, particularly crown rot, which may affect the whole carton and promote uneven fruit ripening. Because of weak pedicels, the fingers of some cultivar also fall from the hand during ripening, exposing the pulp (Nakasone and Paull, 1999).

Transport

Banana bunches after harvest, are transported to the packing shed on padded trailers or on an overhead cable system. Dehanding can be performed in the field, with the hands transported to the packing shed on padded trailers. Care is essential in these steps to avoid any mechanical injury that would reduce fruit quality (Nakasone and Paull, 1999). Dehanding plantains in the field and the use of plastic forms is recommended to protect bunches of plantain during harvesting and transportation to the packaging site in the same manner as exportation from industrial plantations. This reduces mechanical damage and avoids reduction of fruit quality of plantains for exportation (Chang et al., 2000). The utility of this packaging is not obvious in many producing countries as peasants and intermediate wholesalers are accustomed to bunch. In addition, they are not prepared to bear additional costs or extra investment to buy plastic cages for local sales but the loss from mechanically damaged fruits was far higher than the cost of plastic cages (Chang et al., 2000).

In most countries, markets located in the production area, the bunches of plantains bought from the villages are piled up on one another then loaded in bulk in trucks or vans for travel to big distribution and consumption centers situated at times hundreds of kilometers away. The bunches are piled up to maximize loading and to expedite transportation. They are unloaded without caution at the destination. These different modes of packaging and transportation expose the fruits to damage and low market quality (Dadzie, 1998; Chang et al., 2000).

In Ethiopian, large trucks are used to transport unripe banana from major growing areas to big cities. Bunches are piled on the truck loosely and then covered with banana leaf. Due to the distance from the growing area to the cities, fruits may stay a day or two and crating or boxing is not practiced, which results in mechanical damage which is not clearly visible on unripe fruit. However, on ripening it may greatly reduce the market value (Seifu, 1999).

While transporting banana for export purpose, boxing was experimented in the previous times but now abandoned because of various types of spoilage. Modern means of combating the organisms that cause such problems, as well as better systems of handling and transport, quality control, and good container design, have made carton packing not only feasible but necessary. First, the hands are graded for size and quality and then packed in layers in special ventilated cartons with plastic padding to minimize bruising (Morton, 1987). There are several advantages of boxing over naked bunch transport. Transport of hands in boxes has compelled growers to produce particular size of bunch. This has also avoided more handling and export of waste material (Salunkhe and Kadam, 1995).

PACKAGING AND STORAGE

Controlled Atmospheric Storage

Controlled atmosphere (CA) storage is a technique for maintaining the process quality in an environment that differs from air with respect to the proportion of O_2 and CO_2 (Abdullah et al., 1990). Wills et al. (1998) describe CA storage as the precise control of oxygen and carbon dioxide concentrating usually with a decrease in oxygen and increase in the carbon dioxide to extend the produce storage life. The respiration rate of fresh produce is slowed down with a decrease in oxygen of fresh produce. It has previously been reported by Salunkhe and Desai (1984) as cited by Ahmad et al. (2006) that controlled atmosphere with high CO_2 inhibits the breakdown of peptic substances, retains fruit texture and remains firmer for a longer period. In current research this effect was most dominant at reduced O_2 levels but effect of low O_2 cannot be separated from the effect of increased CO_2. However, firmer ripe fruit is considered one of the benefits of CA storage to reduce the mechanical damage, avoid fungal infection and to increase the shelf life of ripe banana fruit.

CA storage condition inhibits ethylene production and retards the rate of banana ripening (Quazi and Freebrain, 1971). However, care has to be taken because low oxygen can cause fermentation/ anaerobic respiration. Therefore, it is necessary to formulate different combinations to delay ripening of banana fruit, while allowing them to subsequently ripen to good quality. Liu (1976) reported that pretreated bananas (ethylene treated) stored for 28 days in 1% oxygen at 14°C remained green and firm until the end of storage, but started to ripen almost immediately after these were placed at 21°C in air without addition of ethylene treatment. Acedo and Bautista (1993) reported that fruit can be successfully stored under low oxygen conditions without raising the carbon dioxide in the atmosphere. According to the study of Ahmad et al. (2006) unripe fruit did not synthesize significant amount of ethylene at 18°C unless the oxygen level was above 7.5 to 8.0. Storage in reduced oxygen and increased carbon dioxide reduced the respiratory activity and slowed down the ripening processes. Nair et al. (1992) observed that oxygen and carbon dioxide quickly equilibrated when bananas were removed from lower oxygen to normal air.

Generally, the application of controlled atmosphere storage has a considerable significance in the proper shipment, storage and ripening of banana as a result, respiretion rate of the produce slowed down as a consequence processes associated with ripening were slowed (Thompson, 2001).

Modified Atmospheric Storage

Due to the complexities involved with produce, that is, varying respiration rates which are product and temperature dependent, different optimal storage temperatures for each commodity, water absorption, by-products, and so on, many considerations are involved in choosing an acceptable packaging technology. One of the areas of research that has shown promise, and had success, is that of modified atmosphere packaging (MAP). This technique involves either actively or passively controlling or modifying the atmosphere surrounding the product within a package made of various types and/ or combinations of films (Farber et al., 2003).

A modified atmosphere can be defined as one that is created by altering the normal composition of air (78% nitrogen, 21% oxygen, 0.03% carbon dioxide and traces of noble gases) to provide an optimum atmosphere for increasing the storage length and quality of food/produce. This can be achieved by using controlled atmosphere storage (CAS) and/or active or passive MAP. Under controlled atmospheric conditions, the atmosphere is modified from that of the ambient atmosphere, and these conditions are maintained throughout storage. MAP uses the same principles as CAS; however, it is used on smaller quantities of produce and the atmosphere is only initially modified. Active modification occurs by the displacement of gases in the package, which are then replaced by a desired mixture of gases, while passive modification occurs when the product is packaged using a selected film type, and a desired atmosphere develops naturally as a consequence of the products' respiration and the diffusion of gases through the film (Farber et al., 2003).

Development of an extended shelf life of banana process should be possible by applying treatments of MAP storage. Once ripening was triggered by not exposing the bananas to large concentrations of ethylene as is normally done, MAP storage conditions decreased respiretion and extended ripening of the bananas up to at least one month (Basel et al., 2002). According to the study of Hassan (2004), the development of modified atmosphere technology has allowed the bananas to be stored for more than eight weeks in combination with refrigeration at 14°C.

Modifications of the storage atmosphere, along with their effects on the physiological life of the produce, can also be advantageous in retarding postharvest disease development. The effect of atmosphere modification on postharvest disease development can be direct-by suppressing various stages of the pathogen growth and its enzymatic activity, or indirect-by maintaining the resistance of the host to infection as a result of keeping it in a superior physiological condition (Irtwange, 2006).

Unripe bananas subjected to ripening in polyethylene bags of 150 gauge thickness at ambient temperature had significantly lower values of pulp: peel ratio, moisture total soluble solid, total sugar, titrable acidity, and total yellow pigments and higher values of alcohol insoluble substances , ascorbic acid, starch, and total chlorophyll, indicating that the ripening in these fruits was retarded to the extent of about 6 to 7 days, followed by those fruits stored in polyethylene bags of 200 gauge with 15 holes per bag (Salunkhe and Kadam, 1995).

Chamara et al. (2000) reported that packaging of a banana cultivar called 'Kolikuttu ' as individual hands in low density polyethylene (LDPE) bags with a wrapped ethylene absorber could be recommended to extend their shelf life at ambient temperature. This could be of considerable economic importance in countries where cold storage is not readily available or expensive.

The benefits of film packaging include easy to handle (consumer package); protection from injuries; reduction of water loss, shrinkage, wilting; reduction of decay by modified atmosphere, reduction of physiological disorders (chilling injury), retardation of ripening and senescence processes and control of insect in some commodities. Harmful effects of film packaging include enhancement of decay due to excess humidity; initiation and/or aggravation of physiological disorders, irregular ripening in improper concentrations of CO_2/O_2; off-flavors and off-odors and increased susceptibility to decay (Irtwange, 2006).

Packaging can influence temperature management and water loss of the produce. To reduce water loss plastic wraps and liners may be used, but the restriction of ventilation so caused can result in problems of low oxygen, high carbon dioxide or an accumulation of water which can be improved by perforation of the films (Burdon, 2001).

Some of permeable films used for packaging of fresh fruits and vegetables are low density polyethylene, linear low density polyethylene, medium density polyethylene, high density polyethylene, polypropylene and polyvinyl chloride with different permeability for gases and water vapor as well (Thompson, 2001; Schlime and Rooney, 1994). The permeability of films to gases (including water vapor) varies with the type of material from which they are made, temperature, in some cases humidity, the accumulation and concentration of the gas and the thickness of the material.

FACTORS AFFECTING STORAGE

Simple methods intended to reduce the desiccation and the evapotranspiration rate of fruits are occasionally used within the traditional distribution channel to maintain a certain level of freshness and an acceptable quality for a number of days. These measures include precautions to limit mechanical damage to fruits, stocking bunches under shades shielded from the sun, and protection of piles of fruits with leaf of banana or bags regularly moistened with water (Kader et al., 1985).

Banana can be stored at a temperature slightly above 13°C and a relative humidity of 85 to 95% for about three weeks. Keeping the fruit in relatively high concentration of CO_2 and low concentration of O_2 can prolong storage life (Manzur et al., 2007). Generally, fruits must be kept at 13.3 to 15.6°C and 80 to 85% relative humidity after removal from storage and during delivery to markets to avoid rapid spoilage.

Temperature

Green bananas are shipped/stored at 13 to 14°C to delay ripening. Lower temperature can lead to chilling injury, whose symptoms include a dull, grey skin colour, poor ripening, and poor conversion of starch to sugars, poor flavor development and susceptibility to decay. Symptom development is dependent on exposure temperature and time and susceptibility depends mostly on cultivar (Nakasone and Paull, 1999).

Temperature beyond 25°C shortened the duration of the preclimacteric and fruit quality is altered because of modification to metabolism during ripening. Above 35°C the development of the peel and the pulp is desynchronized, with softening of the pulp proceeding faster than the colouring of the peel. This results in fruits with a soft pulp but a green peel and this type of fruit is known as cooked/boiled green and a temperature above 48°C, the climacteric in not triggered and ripening is effectively blocked on the other hand, banana fruit blackens at lower temperatures and should not be placed in a refrigerator (Gowen, 1995).

Relative Humidity

The lower the relative humidity, the greater the loss of water and the shorter is the duration of the preclimacteric.

Low humidity brings forward the production of ethylene and the respiratory rise, but its effect on the intensity of ethylene production seems to be dependent on variety. Relative humidity changes in the weight relationship between pulp and peel, in the peel colour, in pulp softening and in the soluble sugar content (Gowen, 1995).

Gas Compositions

The composition of gases in the storage atmosphere can affect the storage life of horticultural products. Changes in the concentration of the respiratory gases, oxygen and carbon dioxide may extend storage life. This is generally used as an adjunct to low temperature storage, but modification of the storage atmosphere can usefully substitute for refrigeration for some commodities (Wills et al., 1998). Storage in reduced oxygen and increased carbon dioxide reduced the respiratory activity and slowed down the ripening processes. Ahmad et al. (2006) observed that oxygen and carbon dioxide quickly equilibrated when bananas were removed from lower oxygen to normal air. According to the report of Knee (1980) chlorophyll breakdown was reduced clearly in reduced oxygen.

According to the study of Shorter et al. (1987), the storage life of banana can be increased five times when they are stored in plastic film (where the gas content stabilized at about 2% O_2 and 5% CO_2) with an ethylene scrubber compared to fruit stored without wraps. If the film was insufficiently permeable, packaging apple bananas in polythene film and storage at 13 to 14°C accumulated levels of CO_2 which could prove toxic to the fruit. Symptoms of CO_2 injury were observed when CO_2 levels were between about 5 and 14%. Injury was characterized by darkening of the skin and softening of the outer pulp while the inner pulp remained hard and astringent.

CONCLUSIONS

In Africa, huge postharvest losses especially of horticultural produce have been reported due to lack of packaging, storage facilities, and poor means of transportation. In spite of these facts, the attention given to reduce this postharvest loss has been low in the past. However, in recent years the postharvest handling of horticultural crops is becoming the area of research and there are some considerable work done in this area on different crops. Banana is one of the most common and widely grown fruit crops. It is a delicate and highly perishable fruit, and the production is subject to poor handling and storage practices and postharvest diseases. As a result, a huge postharvest loss of banana has been reported every time and these postharvest losses are incurred at different stage from harvesting to the final consumption. Traditionally, farmers use teff straw and banana leaf during transportation and storage mainly to avoid mechanical damage during transportation and to avoid water loss during storage. These two materials are better to keep the fruit than keeping in an open air, but there are also other modern packaging materials like polyethylene bags with different permeability to gases and water vapour which improve the shelf life. So, the following aspects should be considered: in order to draw a general recommendation, research should be done on different polyethylene bag packagings and also on the physicochemical properties of the cultivars at different times and in different locations; there is a need for testing the suitability and application of different types of polyethylene bag packagings for banana and other crops; these packaging treatments should be combined with other prepackage treatments like disinfection in order to guarantee the complete post-harvest quality of banana fruit and finally, there is also a need to carry out research not only on the preharvest but also on the postharvest approaches hand in hand to reduce losses and maintain quality

REFERENCES

1. Abbas FMA, Saifullah BR, Yeoh SY, Azhar ME (2012). Comparing physicochemical properties of banana pulp and peel flours prepared from green and ripe fruits. Food Chemistry 129: 312–318.

2. Abdullah H, Lizada MC, Tan SBC, Pantastico EB, Tongdee SC (1990). Storage of banana. pp. 44-63. In: Bananas Handling and Marketing in ASEAN. H. Abdullah and E.B. Pantastico (eds.), Asia.
3. Acedo AL, Bautista OK (1993). Banana fruit response to ethylene at different concentrations of oxygen and carbon dioxide. Asian J. Food Technol. 8(2): 54-60.
4. Ahmad S, Pervez MA, Thompson AK (2007). Effect of harvest maturity and position of hands on the ripening behavior and quality of banana fruit. 127p.
5. Ahmad S, Perviez MA, Thompson AK, Ullah H (2006). Effects of storage of banana in controlled atmosphere before ethylene treatments on its ripening and quality. J. Agric. Res. 44(3):219-229.
6. Arias P, Dankers C, Liu P, Pilkauskas P (2003). The world banana economy. Food and Agriculture Organization of the United Nations. Rome.
7. Awole S, Kebede W, Workneh TS (2011) Postharvest quality and shelf life of some hot pepper varieties. J. Food Sci. Technol. DOI 10.1007/s13197-011-0405-1.
8. Basel RM, Racicot K, Senecal AG (2002). Long shelf life banana storage using MAP storage coupled with postharvest methane cyclopropene (MCP) treatment. Anaheim, California.
9. Beverly RB, Latimer JG, Smittle DA (1992). Postharvest handling: a systems approach. pp. 73-98.In: Preharvest Physiological and cultural effects on postharvest quality of fruits. Academic press INC, London.
10. Bugaud C, Daribo MO, Dubois C (2007). Climatic conditions affect the texture and colour of Cavendish bananas (Grande Naine cultivar). Sci. Hort. 113(3): 238-243.
11. Burdon JN (2001). Postharvest handling of tropical and subtropical fruit for export. pp 1-19: In Mitra, S.K., 2001. Postharvest physiology and storage of tropical and subtropical fruits. CABI publishing. London, UK.
12. Caussiol L (2001). Postharvest quality of conventionally and organically grown banana fruit. An M. Sc. Thesis Presented to Cranfield University, Silsoe. pp 160.

13. Chamara D, Illeperume K, Galappatty PT (2000). Effect of modified atmosphere and ethylene absorbers on extension of storage life of kolikuttu banana at ambient temperature. Sri Lanka. 6 (55): 381-388.
14. Chang PY, Shuji F, Ashrafuzzaman MD, Naoko N, Nobuyuki H (2000). Purification and Characterization of Polyphenol Oxidase from Banana (Musa sapientum L.) Pulp. J. Agric. Food Chem. 48: 2732-2735.
15. Crane JH, Carlos FB, Ian M (2005). Banana growing in the Florida home landscape. Institute of Food and Agricultural Sciences (IFAS), University of Florida, Florida. CSA (2004). Agricultural Statistics 2004. Addis Ababa, Ethiopia. BS publications. pp 209-214.
16. Dadzie BK (1998). Postharvest characteristics of black sigatoka resistant banana, cooking banana and plantain hybrids. International network for banana and plantain (Inibap), Technical Guidelines. Rome, Italy. pp 74-78.
17. Dadzie BK, Orchard JE (1997). Routin postharvest screening of banana/plantain hybrids criteria and methods. International network for banana and plantain (Inibap), Technical Guidelines. Rome, Italy. pp 75-76.
18. Dionisio GA (2012). Revisiting hot water treatments in controlling crown rot of banana cv. Buñgulan. Crop Prot. 33: 59-64.
19. Fallik, E and Aharoni Y (2004). Postharvest physiology, pathology and handling of fresh produce, lecture notes. International research and development course on postharvest biology and technology. The Volcani Centre, Isreal, pp: 30.
20. FAO (1999). Fishery and Agriculture Statistics. Food and Agricultural Organization of the United Nation. Rome, Itali. http://www.fao.org.
21. FAO (2000). Fishery and Agriculture Statistics. Food and Agricultural Organization of the United Nation. Rome, Itali. http://www.fao.org.
22. Farber JN, Harris LJ, Parish ME, Beuchat LR, Suslow TV, Gorney JR, Garrett EH, Busta FF (2003). Microbiological safety of controlled and modified atmosphere packaging of fresh and fresh-cut produce. Comp. Rev. Food Sci. Food saf. 2:142-160.

23. Gowen S (1995). Bananas and plantains. Chapman and Hall, London. 567p.

24. Haidar J, Demisse T (1999). Malnutrition and Xerophthagma in rural communities of Ethiopia. East Afri. Med. J. 10: 590-593.

25. Hassan A (2004). Technologies developed for postharvest handling of Malaysian bananas. International congress on Musa: Harnessing research to improve livelihoods. Horticulture Research Centre, Malaysian Agricultural Research and Development Institute (MARDI), Malaysia. 31p.

26. Hofman PJ, Smith LG (1993). Preharvest effects on postharvest quality of subtropical and tropical fruit. In: Proceedings of an international conference held at Chang Mai, Thailand, 19-23 July. ACIAR Proceedings. No 50.

27. Iqbal M (1996). Type and extent of postharvest losses in horticultural commodities in Pakistan. pp: 33-42. In: Proceedings of National Conference on Postharvest Technology of Horticulture Commodities, 10-12 September 1996, Quetta.

28. Irtwange SV (2006). Application of modified atmosphere packaging and related technology in postharvest handling of fresh fruits and vegetables. Agric. Eng. Inter.: the CIGR Ejournal. Makurdi, Nigeria. 4(8):1-13.

29. Johnson GI, Sharp JL, Mine DL, Oostluyse SA (1997). Postharvest technology and quarantine treatments. In: Litz R.E. (eds). The Mango: Botany, production and uses. Tropical Research and Education Center, USA. pp: 444 -506.

30. Kader AA (1992). Postharvest technology of Horticultural crops. University of California, division of Agriculture and Natural Resources, (quality and safety factors, definition and evaluation for fresh horticultural crops) second edition, publication No 3311. pp. 228-345.

31. Kader AA, Kasmire RF, Michell FG, Reid MS, Sommer NF, Thompson JF (1985). Postharvest technology of horticultural crops. University of California, America. 192p.

32. Kays SJ (1997). Postharvest physiology of perishable plant products. Exon Press Athens, GA. US. 578p.

33. Kebede E (1991). Processing of Horticultural Produce in Ethiopia. Acta Hortic. 270–301.

34. Knee M (1980). Physiological response of apple fruits to oxygen concentrations. Ann. App. Biol. 96(4):243-245.

35. Liu FW (1976). Banana response to low concentrations of ethylene. J. Amer. Soc. Hort. Sci. 101(3):222-224.

36. Lopez A (1998). Bananas and plantain. CORBANA, Corporacion Bananera Nacional. San Jose, C. R.

37. Manzur R, Zia MA, Ali L (2007). How to grow bananas. http://archives.dawn.com/2007/09/03/ebr7.htm.

38. Marriott J, Lancaster PA (1983). Bananas and Plantains. In: Handbook of Tropical Foods. Harvey Jr. TC (Ed), Marcel Dekker, Inc. pp. 85142.

39. Mir N, Beaudry RM (2004). Modified Atmosphere Packaging. <www.ba.ars.usda.gov/hb66/015map.pdf> (accessed 25.01.2012).

40. Morton J (1987). Banana. In: Julia, F., Morton, Miami, FL (eds). Fruits of warm climates. pp. 29–46.

41. Munasque VS, Abdullah H, Gelido MERA, Rohaya MA, Zaipun MZ (1990). Fruit growth and maturation of banana, pp.33-43. In: Hassan, and E.B. Pantastico, 1990. Banana fruit development, postharvest physiology, handling and marketing in ASEAN.

42. Nair H, Tung HF, Wan MW, Rosli M, Ahmad HS, Chang KK (1992). Low oxygen effect and storage Mas banana (Musa, AA group). Acta Hort. 292(21): 209-215.

43. Nakasone HY, Paull RE (1999). Tropical Fruits.CAB Inernational publishing.UK by Biddles Ltd,Guildford and King's Lynn. 445p.

44. Ortiz R, Ferris RSB, Vuylsteke DR (1995). Banana and plantain breeding. pp. 110-145. In: Gowen S. (ed). Bananas and plantains, Natural Resources Institute and Department of Agriculture University of Reading, UK, Chapman & Hall.

45. Pesis E (2004). Respiration and ethylene. International Research and Development on Postharvest Biology and Technology. The Volcani Center, Israel.

46. Picq C, Foure E, Frison EA (1998). Banana and food security. Douala, Cameroon. 798p.

47. Quazi MH, Freebrain HT (1971). The influences of ethylene, oxygen and carbon dioxide on the ripening of bananas. Botanical Gazette. 131(5): 5-14.

48. Raja MB, Khokhar KM (1993). Postharvest horticulture technology and its future prospects. pp:265-277. In: Proceeding of First International Horticulture Seminar, 09-11 January 1992. Pakistan Agricultural Research Council, Islamabad.
49. Salunkhe DK, Desai BB (1984). Postharvest biology of fruits. Vol. 1 CRC Press, Boca, Roaton, Florida. 254p.
50. Salunkhe DK, kadam SS (1995). Banana. pp. 67-90. Hand Book of Fruit Science and Technology. Marcel Dekker, Inc. New York.
51. Samson JA (1991). Tropical fruits. Second eds. Singapore publisher. New York. 335p.
52. Schlime DV, Rooney ML (1994). Packaging of minimally processed fruits and vegetables. pp.135 – 179. In: Willey, R. C. (ed.). 'Minimally Processed Refrigerated Fruits and Vegetables'. Chapman and Hall, New York.
53. Seifu GM (1999). Banana production and utilization in Ethiopia. EARO, Addis Ababa. 58p.
54. Seymour GB (1993). Banana. pp 83-106. IN: Seymour, G.B., J.E. Taylor and G.A. Tucker (eds). Biochemistry of Fruit Ripening. Chapman and Hall, London.
55. Seymour GB, Thompson AK, John P (1987). Inhibition of degreening in the peel of banana ripened at tropical temperatures. Annu. Appl. Biol. 110:145-151.
56. Shewfelt RL, Prussia SE (1993). Postharvest handling, a systems approach. Academic press Limited, USA. pp. 358.
57. Shorter AJ, Scott KJ, Graham D (1987). Controlled atmosphere storage of bananas in bunches at ambient temperature. CSIRO food research, Queensland.
58. Simmonds NW (1959). Bananas. Webster printing service Ltd Bristol. Imperial College of Tropical Agriculture. Trop. Agric. 453-461.
59. Smith NJS (1995). Recognizing and achieving commercial quality in fresh produce. Conference Proceedings, Postharvest quality assessment, Silsoe college, The Association of Applied Biologists and the Institution of Agricultural Engineers, Posth. News info. 6 (3): 28-30.
60. Stover RH, Simmonds NW (1987). Bananas. 3rd ed. Tropical agricultural series. Longman, New York.

61. Tadesse F (1991). Postharvest losses of fruits and vegetables in horticultural state farms. Actra Hortic. 270:261-270.

62. Thompson AK (2001). Controlled atmospheric storage of fruits and vegetables. CAB International Printed in UK Biddles Ltd, Guidford and Kings Lynn, UK. 278p.

63. Tigist A, Workneh TS, Woldetsadik K (2011) Effects of Variety on the Quality of Tomato Stored under Ambient Conditions. J. Food Sci. Technol. DOI: 10.1007/s13197-011-0378-0.

64. Tucker GA (1993). Introduction. IN: Seymour GB, Taylor JE, Tucker GA (eds). Biochemistry of fruit ripening. Chapman and Hall, London. pp 1-52.

65. Turner DW (2001). Bananas and Plantains. In: Mitra, S.k. (ed.). Postharvest physiology and storage of tropical and subtropical fruit. CAB International. London, UK. pp 47-84.

66. Wills R, Mcglasson B, Graham D, Joyce D (1998). Postharvest an Introduction to the physiology and handling of fruits, vegetables and ornamentals. 4thed. UNSW press. Adelaid.

67. Workneh TS, Osthoff G, M.S. Steyn, Engelbrecht GM, Pretorius JC (2010). The effect of pre-harvest treatment, disinfection and storage environment on quality of carrots. J. Food Processing and Preservation 35(3): 331–341.

68. Workneh TS, Osthoff G and Steyn MS (2011). Effects of preharvest treatment, disinfections and storage environment on quality of tomato. J. Food Sci. Technol. DOI: 10.1007/s13197-011-0391-3.

Chapter 9

Role of Alternative Oxidase in Postharvest Stress of Fruit and Vegetables: Chilling Injury

Morteza Soleimani Aghdam

Young Researchers Club, Ahar Branch, Islamic Azad University, Ahar, Iran

ABSTRACT

Chilling injury (CI) imposes a limitation for extending storage of fruit and vegetables. Chilling injury can result from oxidative stress caused by reactive oxygen species (ROS). Alternative oxidase (AOX) as a ROS avoidance genes play pivotal role in defense mechanism against chilling injury derived oxidative stress. Postharvest treatment of fruit and vegetables through methyl salicylate, methyl jasmonate and storage in atmosphere with high O_2 could induce AOX gene expression and led to chilling tolerance. In this review, the main topic is the role of postharvest treatments in tolerance to chilling injury via induction of AOX gene expression and AOX physiological roles in fruit ripening.

INTRODUCTION

Low temperature storage is a postharvest technology used widely to extend the postharvest life of fruit and vegetables and allows the preservation of fruit and vegetables quality after harvest, because low temperatures decrease the speed of cell metabolism and delay plant senescence and fruit ripening (Sevillano et al., 2009). Tropical and subtropical fruit and vegetables are sensitive to low temperatures and suffer chilling injury (CI). The incidence of chilling injury as a physiological disorder limits the application of low temperature storage. Therefore, the impact of chilling injury on the agro-food industry has serious economic consequences (Sevillano et al., 2009). Chilling injury of fruit and vegetables negatively affect their quality and therefore their marketing (Lafuente et al., 2005; Lafuente and Zacarias, 2006; Mulas and Schirra, 2007). To increase the tolerance of fruit and vegetables to chilling injury, postharvest treatments such as signaling molecules: Methyl jasmonate (MeJA) or methyl salicylate (MeSA) (Fung et al., 2004, 2006) or use of storage with high (Zheng et al., 2008) had been developed. Further, the induction of the AOX gene expression by these pretreatments had been shown. Also, there is a direct correlation between the induction of AOX gene expression and the increase tolerance to chilling injury. In the present review, the literature in which the role of AOX had been studied in relation to the use of treatments for harvested fruit and vegetables was analyzed. Besides, the role of AOX in the fruit ripening and their biochemical mode of action were also discussed.

CHILLING INJURY IN HORTICULTURAL CROPS

Tropical and subtropical fruit and vegetables are sensitive to low temperatures and suffer from chilling injury. The occurrence of chilling injury limits the use of cold storage. Therefore, the impact of chilling injury on the horticultural industry has serious economic consequences. Understanding the physiological mechanisms responsible the activation and development of this physiological disorder would allow the design of strategies to avoid or delay CI appearance (Sevillano et al., 2009).

Cell membranes are primary sites for development of CI (Rui et al., 2010). Phase transitions from a flexible liquid crystalline to a solid gel structure occur in membrane of chilled tissue (Lyons, 1973). The fatty acid composition of membrane lipids can determine the existence of a temperature induced phase transition (Martin et al., 1976). In order to explain the impacts of CI on cell membranes and its manifestations at the molecular level when CI occurs, Lyons (1973) proposed the so-called 'membrane theory'. The membrane theory of CI suggests that the immediate effect of low temperatures is a global increase in the microviscosity of the membrane matrix due to a diminution of random rotation or folding movements of the aliphatic chains of fatty acids and to a reduction in the degree of desaturation of these acids. Below a critical value of temperature called the transition temperature, this event leads to a reorganization of the membrane lipids in a rigid structure called solid gel. A higher proportion of unsaturated fatty acids provide higher tolerance to low temperature, as has been reported, for example, in banana, pomegranate and loquat fruit (Mirdehghan et al., 2007; Promyou et al., 2008; Cao et al., 2009). Maintenance of membrane integrity at low temperature has been reported to be important in the resistance to chilling temperature (Wonsheree et al., 2009). As indicators of membrane damage, electrolyte leakage and malondialdehyde (MDA) content are generally considered to be indirect measurements of membrane integrity and can reflect the occurrence of CI and loss of membrane integrity (Shewfelt and Purvis, 1995). Lipolytic cascade in membrane lipids deterioration during senescence and CI was achieved by the concerted activities of a variety of membranous lipolytic enzymes such as phospholipase D (PLD) and lipoxygenase (LOX) (Pinhero et al., 1998). LOX and PLD catalyze peroxidation of polyunsaturated fatty acids and are believed to be major contributors to membrane damage and thus CI in plant (Pinhero et al., 1998; Wang, 2001). Mao et al. (2007) showed that the development of CI in cucumber fruit was accompanied by increase in PLD and LOX activities when exposed to chilling stress, and that the enhanced tolerance to CI by heat treatment was related to the reduction in activities of both enzymes. Rui et al. (2010) reported that the heat treatment increased LOX and PLD activity in response to chilling stress in loquat fruit and the reduction of internal browning (IB), main symptom of chilling injury in loquat fruit by heat treatment was associated with the reduction of PLD and LOX activities. This result suggested that these two enzymes might be associated with the initiation of CI by being involved in membrane deterioration and signalling pathway in response to chilling stress. Changes in membrane structure and composition are considered as the primary event of CI and lead to a loss of permeability control and

metabolic dysfunctioning (Lyons, 1973). It was reported that membrane lipids from chilling resistant plant species showed higher content of unsaturated fatty acids than did sensitive species (Wonsheree et al., 2009). Heat treatment induced acclimatization of pomegranate and banana fruit to low temperature and thus reduced CI by maintaining higher unsaturated/saturated fatty acid ratio (Mirdehghan et al., 2007; Promyou et al., 2008). Cao et al. (2009) showed that the decrease of lipid unsaturation was involved in the induction of CI in loquat fruit. The higher unsaturated/ saturated fatty acid ratio contributed to the reduced CI in methyl jasmonate treated fruit. Rui et al. (2010) suggested that the loss of membrane integrity, decrease in membrane lipid unsaturation and increases in PLD and LOX activities may be involved in the development of IB and CI in loquat fruit. The reduction of IB in chilled loquat fruit by heat treatment might be due to maintenance of membrane integrity, higher unsaturated/saturated fatty acid ratio and reduced PLD and LOX activities.

Apart from the direct effect of low temperatures on the molecular organization of membrane lipids, the loss of integrity of the membrane itself is boosted by oxidative processes, since low-temperature stress increases the levels of reactive oxygen species (ROS) (Sevillano et al., 2009). Defense against oxidative stress consists of two lines: The first line of defense is termed ROS scavenging genes which includes superoxide dismutase (SOD), catalase (CAT), the ascorbate/glutathione (AsA-GSH) cycle, the glutathione peroxidase and thioredoxin system (Møller, 2001). SOD catalyses the dismutation of $O_2^{\bullet-}$ to H_2O_2 and CAT scavenge H_2O_2 to form oxygen and water. The AsA-GSH cycle is also an important mechanism in the removal of ROS in plants. Its activation seems to produce a positive effect by inhibiting the development of CI. Sato et al. (2001) showed that the induction of APX activity by means of a heat treatment is a key element in the protection of rice against a later exposure to low temperatures. The GSH content and GR activity are also significant in plants showing tolerance to low temperatures. Cold-tolerant genotypes of tomato (Walker aand Mckersie, 1993) accumulate more GSH, and during cooling, their GR activity is higher than that of sensitive genotypes. Moreover, the rise in GSH content and

GR activity in cold-sensitive genotypes of corn causes a diminution of CI (Kocsy et al, 2001).

The second line of defense is termed ROS avoidance genes which include alternative oxidase (AOX). The plant AOX pathway branches from the main respiratory electron transport chain, bypasses the final steps of the cytochrome respiratory pathway and catalyses the oxidation of ubiquinol. It was suggested that by maintaining the flow of mitochondrial electrons, AOX maintained activation of NAD(P)H dehydrogenase and proton-pumping NADH dehydrogenase (Møller, 2001) and helps in generation of sufficient ATP required for the rapid adaptation and the maintenance of plant growth rate homeostasis (Moore et al., 2002; Hansen et al., 2002). In this process, AOX is involved in the reduction of ROS by preventing electrons from reducing O_2 to $O_2^{\bullet -}$ and thus reduces the level of O_2 in the mitochondria (Møller, 2001; Wagner and KRAB, 1995; Mittler, 2002). The concept of AOX acting as an antioxidant enzyme has been shown in isolated bell pepper mitochondria (Purvis, 1995) and in intact tobacco cells (Maxwell et al., 1999). It was proposed that the alternative respiratory pathway mediated chilling injury by keeping the production of ROS in balance with the levels of antioxidants and active oxygen scavenging enzyme systems (Purvis, 1995).

AOX FUNCTIONS IN FRUIT AND VEGETABLES

Two main hypotheses for roles of AOX activity in plants suggested that: (i) AOX activity allows continued TCA cycle carbon flow to occur under conditions when the cytochrome pathway is inhibited or restricted by the availability of ADP (Bahr and Bonner, 1973; Vanlerberghe and Mcintosh, 1994). When mitochondrial electron transport via the cytochrome pathway is inhibited or restricted, it results in the inhibition of TCA cycle and glycolysis. Reducing power (NAD (P) H) accumulates, biosynthesis stops and aerobic fermentation takes place which may lead to tissue damage by ethanol production (Vanlerberghe et al., 1995). An alternative electron transport pathway to oxygen could prevent redirection of carbon metabolism to the fermentative pathways and allow activity of TCA cycle to continue. Palmer (1976) suggested that this was probably the primary function of the AOX. Such a function was later modified by Lambers (1985), who suggested that the alternative pathway functions as an "energy overflow". It was proposed that the alternative pathway became engaged when the cytochrome pathway was saturated (Bahr and Bonner, 1973) such as the one that occur in the presence of excess sugars. (ii) The AOX has a role in the defence mechanism against oxidative stress such as chilling injury (Sevillano et al., 2009). Mittler (2002) proposed that the AOX reduced ROS production by means of two mechanisms: preventing the reduction of O_2 to superoxide radical and reducing the amount of O_2, the substrate for the formation of ROS in cell organelles.

There are two energy dissipating systems in plants: The redox energy dissipating pathway (AOX) and the proton electrochemical gradient (H+) energy dissipating pathway or plant uncoupling mitochondrial protein (PUMP) which leads to the same final effect such as a decrease in ATP synthesis and an increase in heat production (Sluse and Jarmuszkiewicz, 2000). Linoleic acid, an abundant free fatty acid (FFA) in plants which activates PUMP, strongly inhibits cyanide resistant respiration mediated by AOX. Sluse and Jarmuszkiewicz (2000) reported that the AOX and PUMP acts sequentially during postharvest ripening of tomato fruit, AOX activity decreases in early post-growing stages and PUMP activity is decreased in late ripening stages. Thermogenesis occurs during fruit ripening (Cruz-Hernandez and Gomez-Lim, 1995; Kumar et al., 1990; Kumar and Sinha, 1992) and FFA concentration increases during the post-growing stage (Rouet-Mayer et al., 1995). Sluse and Jarmuszkiewicz (2000) indicated a clear regulation of AOX activity through a decrease in protein expression during tomato fruit ripening, and a decrease in PUMP protein expression only after the yellow stage that parallels PUMP activity. Sluse and Jarmuszkiewicz (2000) suggested that AOX and PUMP acts sequentially. AOX would be active mainly during the growing period, thereby providing a safety balance between redox potential, phosphate potential and biosynthesis demand, whereas PUMP would start working in the post-growing stage when the FFA concentration increases, thereby providing a mechanism for heat generation via a decrease in the efficiency of oxidative phosphorylation in parallel with the termination of biosynthetic processes.

Considine et al. (2001) reported that the expression of PUMP genes is strongly enhanced during fruit ripening, suggesting a developmental stage specific regulation of PUMP. The increase in mRNA levels after the climacteric ripening phase directly correlates with the increase of mango PUMP content. In tomato, the studies on fruit ripening show that PUMP expression adapts to physiological conditions. Holtzapffel et al. (2002) showed an increase in transcript and protein levels in the later stages of on-vine fruit ripening and Almeida et al. (1999) reported that the activity of PUMP decreases during the later stages of postharvest fruit ripening. Almeida et al. (1999) also reported a higher amount of AOX in tomato fruit in postharvest ripening. However, FA concentrations also increase during the post growing stage. The PUMP and AOX pathways could therefore operate with different efficiencies under distinct physiological conditions. AOX would be active mainly during high biosynthetic activities in early stages of tomato fruit ripening, whereas PUMP would be functionally silent. With increasing FFA concentration in post growth stages, as in fruit ripening but perhaps also in senescence and flowering, PUMP activity could reach maximum velocity while AOX activity is switched off (Jarmuszkiewicz et al., 2001).

POSTHARVEST TREATMENTS, AOX GENE EXPRESSION AND CHILLING TOLERANCE IN FRUIT AND VEGETABLES

Although chilling injury can be prevented by maintaining the commodity at temperatures above the critical threshold, these temperatures can significantly reduce the product's shelf life (Troncoso-Rojas and Tiznado-Hernandez, 2006). Fruit and vegetables that are susceptible to chilling injury have a short storage life in low temperature storage. Low temperature storage cannot be used to slow losses and pathogen growth in this commodity, which has serious economic consequence on the agro-industry.

Valero and Serrano (2010) reported that the potential symptoms of chilling injury in fruit and vegetables are surface lesions, pitting, sunken and discoloration area, water soaking of tissue, water loss, desiccation, shriveling, internal browning, mealiness and browning in the flesh, tissue breakdown, failure of fruit to ripen or uneven or slow ripening, accelerated senescence and ethylene production, shortened storage or shelf life, loss of flavor, loss of growth or sprouting capability, wilting and increased decay due to leakage of cell metabolites.

Due to consumer concerns about the presence of chemicals in fruit and vegetables, the use of environ-mentally friendly technologies such as salicylic and jasmonic acids and their methyl esters to avoid the development of chilling injury have been considered in many studies. The exposure of fruit and vegetables to these signaling molecules is used as a strategy to protect the commodity from the subsequent stress represented by the cold storage. From a biochemical point of view, evidence that links this protective effect with the expression of AOX has been found. The present review discusses the application of postharvest treatments and the link between treatment application and AOX gene expression.

SIGNALING MOLECULES: SALICYLATES AND JASMONATES

The increasing demand for consumption of fresh fruit and vegetables, along with more restriction on the use of synthetic chemicals to preserve produce quality and reduce CI, has encouraged scientific research to develop new technologies based on natural product such as salicylic acid and methyl salicylate (MeSA), and jasmonic acid and methyl jasmonate (MeJA).

SA and MeSA are endogenous signal molecules, playing pivotal roles in regulating stress responses and plant development. Recently, SA has received a particular attention because it is a key signal molecule for expression of multiple modes of plant stress resistance. Although, the focus has been mainly on the roles of SA on biotic stresses, several studies support major roles of salicylates in modulating plant response to several abiotic stresses, such as UV light, drought, salinity, chilling stress and heat shock. Salicylates delay the ripening of fruits, probably through inhibition of ethylene biosynthesis or action, and maintain postharvest quality (Asghari and Aghdam, 2010). Although, there are many methods to reduce CI in various fruit and vegetables, SA and MeSA treatments are inexpensive, easy to apply and can be used on various horticultural crops (Asghari and Aghdam, 2010). SA has been shown to induce expression of AOX and increase the antioxidant capacity of the cells. Increases in AOX transcript levels using SA and MeSA before cold treatment reduces the incidence of CI in freshly green bell peppers (Fung et al., 2004). Lipid peroxidation is closely tied to CI in plant cells and this leads to MDA accumulation. MDA accumulation is prevented by SA treatment (Asghari and Aghdam, 2010). SA as a signal triggers the induction of cyanide resistance respiration in plant cells by affecting the AOX enzyme activity (Raskin et al., 1989). In fruit and vegetables, SA affects AOX activity leading to decrease in the harmful effects of different post-harvest oxidative stresses such as chilling injury, prevents fermentation and maintains low respiration rates and decreases fruit ripening and senescence rates. Respiration of harvested crops is highly dependent on ethylene production and activity and any factor increasing the production and activity of ethylene leads to increases in respiration and consequently increases the senescence rate. Effect of SA in decreasing the respiration rate is mainly due to its negative effects on ACC, ACO, PG, PME, cellulase and antioxidant enzymes leading to decrease in ethylene production and action (Asghari and Aghdam, 2010).

MeJA treatment also regulates diverse processes such as skin color development by promoting β-carotene synthesis and chlorophyll degradation, CI and ion leakage. MeJA treatment reduces the development of CI symptoms in a wide range of fruit and vegetables, including mango (González-Aguilar et al., 2000), guava (González-Aguilar et al., 2004), tomato (Ding et al., 2001) and loquat (Cao et al., 2010). Fruit and vegetables treated with MeJA have higher sugar, organic acid and vitamin C levels. MeJA stimulate ethylene biosynthesis by increasing ACS and ACO activity, which in turn enhances fruit ripening (González-Aguilar et al., 2006). MeJA treatment inhibits the gray mould rot caused by *Botrytis cinerea* and reduces decay caused by green mould *Penicillium digitatum* (Gonzalez-Aguilar et al., 2006). Recently, it was reported that MeJA induced the expression of pathogen related proteins (PRs) and AOX genes, increase the transcript accumulation of heat shock proteins (HSPs), and enhanced antioxidant system activity. These findings help to explain the mode of action of MeJA in increasing chilling tolerance in fruit and vegetables (Gonzalez-Aguilar et al., 2006).

Fung et al. (2004) reported that the treatment of pepper fruit with MeSA or MeJA vapors increased preferentially the transcription levels of AOX. These authors showed that the overnight treatment with MeSA or MeJA vapors increased transcript levels of AOX (1.5 kb) even at room temperature of 25°C, whereas no change was observed with untreated control. In addition to the expected 1.5 kb AOX transcript, RNA gel blot analysis by Fung et al. (2004) revealed an extra 3.5 kb transcript which was induced only at 0°C. At 0°C, both AOX transcripts reached maximal levels firstly in MeSA treated fruit, secondly in MeJA treated fruit and lastly in controls.

Fung et al. (2004) showed that the MeSA and MeJA vapors increased resistance against chilling injury in freshly harvested green bell pepper. They also showed that the increase in AOX transcript levels by MeJA or MeSA before cold treatment was correlated with reduced incidence of chilling injury in green bell pepper. AOX transcript, protein levels and activity were shown to respond to multiple developmental and environmental triggers (Considine et al., 2002). For example, AOX transcript and protein can be increased by low temperature (Ito et al., 1997; Djajanegara et al., 1999). Mutated AOX gene was linked to quantitative trait loci (QTL) for low temperature tolerance in a rice cultivar (Ribas-Carbo et al., 2000).

Fung et al. (2004) reported that the expression of AOX family 1 gene(s) in pepper was tightly associated with low temperature stress, suggesting that family 1 gene play a role in stress-induced conditions in plants. Within 6 h of rewarming, AOX transcripts (3.5 and 1.6 kb) in pepper fruit are depleted suggesting the presence of a high turnover rate for AOX transcripts. Low levels of CaAOX1 transcripts are detected in fruit stored at 20°C and freshly harvested. This suggests that the CaAOX1 transcript is constitutively expressed throughout fruit development and corresponds to basal levels necessary for alternative respiration (Fung et al., 2004).

Fung et al. (2004) showed that the CaAOX1 transcripts (1.5 kb) reach high levels within 2 days at 5°C but fruit at 0°C takes up to 7 days to attain the same level. Purvis (2001) reported that the higher AOX protein levels were found in mitochondria from pepper fruit stored at 4°C than at 1°C. Fung et al. (2004) suggested that the delay of CaAOX1 transcript accumulation at 0°C can be overcome by application of MeJA or MeSA vapors. Both MeSA and MeJA reduce short-term external surface pitting and long-term internal seed blackening symptoms in pepper fruit. Though, MeSA induced higher and earlier expression of CaAOX1 transcripts, MeJA (10−4 M) was more effective than MeSA (10−4 M) at alleviating chilling injury. It is possible that the beneficial effects for chilling resistance from enhancement of alternative oxidase might be compromised by the death-inducing properties of MeSA (Alvarez, 2000).

Fatty acid desaturase (18:1 FAD) gene for the production of polyunsaturated membrane lipid (Miquel et al. 1993), the catalase and alternative oxidase (AOX) genes in the oxidative stress defense mechanism (Kerdnaimongkol and Woodson, 1999; Abe et al., 2002), cold regulated (COR) genes such as dehydrins (Thomashow, 1998) and CBF1 genes, a member of AP2/EREBP transcription factor family (ZHAO et al., 2009) are mechanisms correlated with the acquisition of chilling tolerance.

AOX action maintains the flow of mitochondrial electrons and avoids production of ROS (Purvis, 1997; Maxwell et al., 1999; Møller, 2001) which was thought to be the main factor which resulted in CI (Purvis et al., 1995). MeSA and MeJA induced CaAOX1 gene expression is correlated with resistance to CI in peppers (Fung et al., 2004). One unexplained phenomenon of CaAOX1 expression was the detection of two CaAOX transcripts (1.5 and 3.5 kb) at low temperature. A longer transcript was also detected in rice and suggested to be the unprocessed transcript of AOX (Ito et al., 1997). Family 1 AOX genes typically occur as multigene families in plants (Ito et al., 1997). For example, closely related LeAOX1a and 1b gene transcripts are expressed in chilled tomatoes (Holtzapffel et al., 2003). By using a yeast expression system, the 1b protein was shown to be functional and had altered regulatory properties in comparison to 1a. The 1b protein was suggested to be a less regulated form of AOX that remains active under stress conditions (Holtzapffel et al., 2003). Overnight exposure to MeSA or MeJA was effective at increasing resistance to CI in tomatoes at the mature green maturity stage (Ding et al., 2002) and in peppers (Fung et al., 2004).

Fung et al. (2006) treated pink tomato fruit overnight with MeJA or MeSA at a concentration of 10-4 M before cold storage at 0°C. CI index was recorded based on overall fruit appearance by taking into account the degree of shrivel, pitting, decay and fruit color. These authors reported that the significant differences in CI index were found among treated and untreated pink tomato fruit only after 3 weeks of storage and only with MeSA treatment. For MeJA treated fruit, no differences were observed in untreated fruit except for less decay. For MeSA treated fruit, CI index was significantly lower in terms of the degree of shrivel and decay. Red color development was inhibited in both untreated and MeJA treated fruit and was characterized by the development of pale orange and yellow pigmentation. In summary, MeSA (but not MeJA) efficiently delayed CI in tomatoes treated at the pink maturity stage (Fung et al., 2006).

RNA gel blot analysis carried out by Fung et al. (2006) indicated that LeAOX1a mRNA abundance increased during the 24 h exposure to exogenous MeSA and also during the first day at 0°C in both MeJA and MeSA treated fruit. In addition to the 1.6 kb LeAOX1a transcript, a 3 kb transcript was detected when fruits were transferred at 1°C. Both 1.6 and 3 kb LeAOX1a transcript levels remained high in treated and untreated fruit throughout the first 5 days of 0°C storage. The expression profile of LeAOX Family 1 transcripts in fruit appeared to be determined almost entirely by LeAOX1a expression. The transcript levels of the LeAOX Family 2 gene appeared to be increased by MeSA and MeJA during days 1 and 2, and thereafter remained unchanged during cold storage. A 3 kb transcript was also detected for the LeAOX Family 2 gene during 0°C storage.

Fung et al. (2006) concluded that the CI index and degree of severity (shrivel and decay) correlates with the expression pattern of the LeAOX1a and LeAOX2 genes in MeSA treated fruit before and during the first day of cold storage. The three closely related LeAOX Family 1 genes are expressed at different levels in fruit and LeAOX1a is responsible for the vast majority of LeAOX Family 1 expression in fruit during cold storage and accumulation of its mRNA correlates with chilling resistance. MeJA and MeSA treatments induced chilling resistance in mature green tomatoes (Ding et al., 2002) and green bell peppers (Fung et al., 2004). However, MeJA is not as effective as MeSA in alleviating CI of pink maturity stage tomatoes. Fung et al. (2006) results confirms previous work demonstrating that tomato fruit at different maturity stages respond differently to plant growth regulator treatment (Ding and Wang, 2003). Also, chilling tolerance of maize, cucumber and rice seedling leaves and roots was shown to be differentially affected by salicylic acid (Kang and Saltveit, 2002). The delayed induction of LeAOX1a and LeAOX2 transcript levels in MeJA treated fruit may provide one explanation for the lower degree of shrivel found among chilling resistant MeSA treated tomato fruit but not in MeJA or control fruit. Among the AOX gene family members, LeAOX1 transcripts accumulate to the highest levels during cold storage and LeAOX1a mRNA abundance is higher in comparison to that of the 1b and 1c genes in fruit throughout the course of the treatment.

HIGH OXYGEN STORAGE

Storage of horticultural crops under superatmospheric O_2 concentration may stimulate, have no effect, or reduce ethylene production and rates of respiration, depending on the commodity, maturity and ripeness stage, O_2 concentration, storage time and temperature, and concentrations of CO_2 and ethylene present in the atmosphere. Ripening of mature-green climacteric fruits was slightly enhanced by exposure to 30 to 80 kPa O_2, but levels above 80 kPa retarded their ripening and caused O_2 toxicity disorders on fruits. High O_2 concentrations enhance the effects of ethylene on fresh fruits and vegetables, including ripening, senescence and ethylene-induced physiological disorders such as bitterness of carrots and russet spotting on lettuce. While superatmospheric O_2 concentrations inhibit the growth of bacteria and fungi, they are much more effective if combined with elevated (15 to 20 kPa) CO_2, which is a fungistatic gas (Kader and Ben-Yehoshua, 2000). AOX gene expression is enhanced by elevated O_2 atmospheres (Zheng et al., 2008). Zheng et al. (2008) reported that the transcript levels of antioxidative genes including SOD, APX and CAT which were relatively constant during storage at 5°C with high oxygen treatment in freshly harvested zucchini squash. However, the expressions of AOX were induced slightly in squash treated with 60 and 100% oxygen for 3 days when compared with the control squash. These increases in AOX transcript levels were correlated with the increased chilling resistance in the treated squash. The ROS scavenging enzyme activities including SOD, APX, CAT and POD in treated zucchini squash were also higher than those in the control for the first 3 days at 5°C. Transcript levels of AOX increased substantially between 3 and 6 days in all treatments, suggesting the involvement of alternative respiratory pathway during chilling stress. All the enzyme activities in 100% oxygen treated squash started to decline after 6 or 9 days of cold storage to a level comparable or lower than those of the control. These declines were correlated to the loss of chilling resistance in the 100% oxygen treated tissue as indicated in the chilling injury index. However, squash treated with 60% oxygen maintained elevated levels of all enzyme activities except POD and sustained the least chilling injury throughout the 15 days of storage at 5°C. The oxygen radical absorbance capacity (ORAC) values and total phenol content remained high in squash treated with 60 and 100% oxygen for the first 9 days, then their levels in the 100% oxygen treated samples declined sharply while those in the 60% oxygen treated samples maintained elevated, indicating that both ORAC activity and phenolic content may also contribute to the resistance of tissue against chilling injury. The 100% oxygen treated squash showed the lowest respiration rate and 60% oxygen treated samples had the lowest ethylene production. Zheng et al. (2008) results may be an indication of the low chilling injury in the high oxygen treated squash. Zheng et al. (2008) results showed that the ORAC and total phenolic levels in zucchini skin peels were both induced by cold storage and further enhanced by 60% oxygen storage. The enhanced antioxidative enzyme activities and the overall antioxidative capacity (ORAC and phenolic levels) seem to correlate with the reduced chilling injury.

In peppers, treatment with MeJA and MeSA reduced chilling injury and induced preferentially the transcript level of AOX gene, while no change in transcript levels of ROS scavenging genes were detected during the early cold storage at 0°C (Fung et al., 2004). Zheng et al. (2008) showed that the elevated level of oxygen was effective in reducing chilling injury in zucchini. Similarly, Zheng et al. (2008) observed slight induction of only AOX gene transcript during early chilling period at day 3 when no difference in chilling injury symptom was visible among treated and control squash. The detectable differences in AOX transcript levels among treated and control zucchini and peppers (Fung et al., 2004) suggested possible correlation between AOX gene expression level and chilling injury. Zheng et al. (2008) results are consistent with the idea that global reprogramming of metabolism occurs during low temperature stress (Sung et al., 2003; Graya and Heath, 2005; Hannah et al., 2005). Zheng et al. (2008) observed that cold storage resulted in inhibition of primary metabolic pathways of respiration and induction of AOX transcript of alternative respiratory chain. This is also accompanied by the antioxidant system as demonstrated by their enzyme activities and by the ORAC assay that presumably function to provide immediate relieve to the adverse redox status. On the other hand, the plant system quickly remobilize its primary metabolites into various secondary metabolite biosynthetic pathway (Kaplan et al., 2004; Renaut et al., 2005) as also shown by Zheng et al. (2008) data from the total phenolic compound that increased by 30% during cold storage.

CONCLUSIONS

AOX is a best candid gene for manipulation of resistance to chilling injury in fruit and vegetables. Postharvest treatments increased the AOX gene expression and can be used in postharvest technology as an important method for reducing chilling injury of fruit and vegetables. The applied technologies for the reduced CI impacts via the increase of AOX gene expression are based mainly on the application of signalling molecules such as SA, MeSA and JA, and MeJA or storage with high O_2. Biotechnology has also offered way to reduce the impact of CI in sensitive fruit and vegetables with the generation of genetically modified plants by the overexpression of AOX genes that act as an antioxidant avoidance gene and controls the ROS production in fruit and vegetables under chilling temperature.

REFERENCES

1. Abe F, Saito K, Miura K, Toriyama K (2002). A single nucleotide polymorphism in the alternative oxidase gene among rice varieties differing in low temperature tolerance. FEBS Lett. 527: 181-185.
2. Almeida AM, Jarmuszkiewicz W, Khomsi H, Arruda P, Vercesi AE, Sluse FE (1999). Cyanide resistant, ATP-synthesis-sustained, and uncoupling-protein-sustained respiration during postharvest ripening of tomato fruit. Plant Physiol. 119: 1323-1329.
3. Asghari M, Aghdam MS (2010). Impact of salicylic acid on postharvest physiology of horticultural crops. Trends Food Sci. Technol. 21: 502-509.

4. Alvarez ME (2000). Salicylic acid in the machinery of hypersensitive cell death and disease resistance. Plant Mol. Biol. 44: 429-442.

5. Bahr JT, Bonner WD (1973). Cyanide-insensitive Respiration II. Control of the alternate pathway. J. Bio Chem. 248: 3446-3450.

6. Cao SF, Zheng YH, Wang KT, Jin P, Rui HJ (2009). Methyl jasmonate reduces chilling injury and enhances antioxidant enzyme activity in postharvest loquat fruit. Food Chem. 115: 1458-1463.

7. Cao S, Zheng Y, Wang K, Rui H, Tang S (2010). Effect of methyl jasmonate on cell wall modifications of loquat fruit in relation to chilling injury after harvest. Food Chem. 118: 641-647.

8. Considine MJ, Daley DO, Whelan J (2001). The expression of alternative oxidase and uncoupling protein during fruit ripening in mango. Plant Physiol. 126: 1619-1629.

9. Considine MJ, Holtzapffel RC, Day DA, Whelan J, Millarm AH (2002). Molecular distinction between alternative oxidase from monocots and dicots. Plant Physiol. 129: 949-953.

10. Cruz-Hernandez A, Gomez-Lim MA (1995). Alternative oxidase from mango (*Mangifera indica*, L) is differentially regulated during fruit ripening. Planta. 197: 569-576.

11. Ding CK, Wang C, Gross K, Smith D (2002). Jasmonate and salicylate induce the expression of pathogenesis related protein genes and increase resistance to chilling injury in tomato fruit. Planta. 214: 895-901.

12. Ding CK, Wang C (2003). The dual effects of methyl salicylate on ripening and expression of ethylene biosynthetic genes in tomato fruit. Plant Sci. 164: 589-596.

13. Ding CK, Wang C, Gross K, Smith D (2001). Reduction of chilling injury and transcript accumulation of heat shock proteins in tomato fruit by methyl jasmonate and methyl salicylate. Plant Sci. 161: 1153-1159.

14. Djajanegara I, Holtzapffel PM, Finnegan PM, Hoefnagel MHN, Berthold DA, Wiskich JT, Day DA (1999). A single amino acid change in plant alternative oxidase alters the specificity of organic acid activation. FEBS Lett. 454: 220-224.

15. Fung RW, Wang CY, Smith DL, Gross KC, Tian M (2004). MeSA and MeJA increase steady-state transcript levels of alternative

oxidase and resistance against chilling injury in sweet peppers (*Capsicum annuum* L.). Plant Sci. 166: 711-719.

16. Fung RW, Wang CY, Smith DL, Gross KC, Tao Y, Tian M (2006). Characterization of alternative oxidase (AOX) gene expression in response to methyl salicylate and methyl jasmonate pre-treatment and low temperature in tomatoes. J. Plant Physiol. 163: 1049-1060.
17. González-Aguilar GA, Fortiz J, Cruz R, Báez CY, Wang CI (2000). Methyl jasmonate reduces chilling injury and maintains postharvest quality of mango fruit. J. Agric. Food Chem. 48: 515-519.
18. González-Aguilar GA, Tiznado-Hernández ME, Wang CY (2006). Physiological and biochemical response of horticultural products to methyl jasmonate. Stew Postharvest Rev. 2: 1-9.
19. González-Aguilar GA, Tiznado-Hernández ME, Zavaleta-Gatica R, Martínez-Téllez MA (2004). Methyl jasmonate treatments reduce chilling injury and activate the defense response of guava fruits. Biochem Biophy Res Commun. 313: 694-701.
20. Graya GR, Heath D (2005). A global reorganization of the metabolome in Arabidopsis during cold acclimation is revealed by metabolic fingerprinting. Physiol Plant. 124: 236-248.
21. Hannah MA, Heyer AG, Hincha DK (2005). A global survey of gene regulation during cold acclimation in Arabidopsis thaliana. PLoS Gen. 1: 179-196.
22. Hansen LD, Church JN, Matheson S, Mccarlie VW, Thygerson T, Criddle RS, Smith BN (2002). Kinetics of plant growth and metabolism. Thermochimica Act. 388: 415-425.
23. Holtzapffel RC, Finnegan PM, Millar AH, Badger MR, Day DA (2002). Mitochondrial protein expression in tomato fruit during on-vine ripening and cold storage. Func. Plant Biol. 29: 827-834.
24. Holtzapffel RC, Castelli J, Finnegan PM, Millar AH, Whelan J, Day DA (2003). A tomato alternative oxidase protein with altered regulatory properties. Bioch. Biophy. Acta. 1606: 153-162.
25. Ito Y, Saisho D, Nakazono M, Trutsumi N, Hirai A (1997). Transcript levels of tandem-arranged alternative oxidase genes in rice are increased by low temperature. Gene. 203: 121-129.

26. Jarmuszkiewicz W, Sluse-Goffart CM, Vercesi AE, Sluse FE (2001). Alternative oxidase and uncoupling protein: thermogenesis versus cell energy balance. Bios Report. 21: 213-22.
27. Kader AA, Ben-Yehoshua S (2000). Effects of superatmospheric oxygen levels on postharvest physiology and quality of fresh fruits and vegetables. Postharvest Biol. Technol. 20: 1-13.
28. Kang HM, Saltveit ME (2002). Chilling tolerance of maize, cucumber and rice seedling leaves and roots are differentially affected by salicylic acid. Physiol Plant. 115: 571-576.
29. Kerdnaimongkol K, Woodson WR (1999). Inhibition of catalase by antisense RNA increases susceptibility to oxidative stress and chilling injury in transgenic tomato plants. J. Am. Soc. Hortic. Sci.124: 330-336.
30. Kocsy G, Galiba G, Brunold C (2001). Role of glutathione in adaptation and signaling during chilling and cold acclimation in plants. Physiol Plant. 113: 158-164.
31. Kumar S, Patil BC, Sinha SK (1990). Cyanide-resistant respiration is involved in temperature rise in ripening mangoes. Biochem Biophy Res Commun. 168: 818-822.
32. Kumar S, Sinha SK (1992). Alternative respiration and heat production in ripening banana fruits (*Musa paradisica* var. Mysore Kadali). J. Exp. Bot. 43: 1639-1642.
33. Lafuente MT, Zacarias L (2006). Postharvest physiological disorders in
34. citrus fruit. Stew Postharvest Rev. 2:1-9.
35. Lafuente MT, Zacarias L, Sala J, Sanchez-Ballesta M, Gosalbes M, Marcos J, González-Candelas L, Lluch Y, Granell A (2005). Understanding the basis of chilling injury in citrus fruit. Acta Hort. 682: 831-842.
36. Lambers H (1985). Encyclopedia of Plant Physiology, New Series (Douce, R. and Day, D. A. eds), Springer-Verlag, Berlin, pp. 418-473.
37. Lyons JM (1973). Chilling injury in plants. Annu. Rev. Plant Physiol. 24:
38. 445-466.

39. Mao L, Pang H, Wang G, Zhu C (2007). Phospholipase D and lipoxygenase activity of cucumber fruit in response to chilling stress. Postharvest Biol. Technol. 44: 42-47.

40. Martin CE, Hiramitsu K, Kitajima Y, Nozawa Y, Skriver L, Thompson GA (1976). Molecular control of membrane properties during temperature acclimation. Fatty acid desaturase regulation of membrane fluidity in acclimating Tetrahymena cells. Biochemistry. 15: 5218-5227.

41. Maxwell DP, Wang Y, Mcintoch L (1999). The alternative Oxidase lowers mitochondrial reactive oxygen production in plant cell. PNAS. 96: 8271-8276.

42. Miquel M, James .Jr. D, Dooner H, Browse J (1993). Arabidopsis requires polyunsaturated lipids for low temperature survival. PNAS. 90: 6208-6212.

43. Mirdehghan SH, Rahemi M, Martinez-Romero D, Guillén F, Valverde JM, Zapata PJ, Serrano M, Valero D (2007). Reduction of pomegranate chilling injury during storage after heat treatment: Role of polyamines. Postharvest Biol. Technol. 44: 19–25.

44. Mittler R (2002). Oxidative stress, antioxidants and stress tolerance. Trends Plant Sci. 7: 405-410.

45. Moller IM (2001). Plant mitochondria and oxidative stress: electron transport, NADPH turnover, and metabolism of reactive oxygen species, Annu. Rev. Plant Physiol., Plant Mol. Biol. 52: 561-591.

46. Moore AL, Albury MS, Crichton PG, Affourtit C (2002). Function of the alternative oxidase: is it still a scavenger? Trends Plant Sci.7: 478-481.

47. Mulas MM, Schirra M (2007). The effect of heat conditioning treatments on the postharvest quality of horticultural crops. Stew Postharvest Rev. 3: 1-6.

48. Palmer JM (1976). The organization and regulation of electron transport in plant mitochondria. Annu. Rev. Plant Physiol. 27: 133-157.

49. Pinhero RG, Paliyath G, Yada RY, Murr DP (1998). Modulation of phospholipase D and lipoxygenase activities during chilling. Relation to chilling tolerance of maize seedlings. Plant Physiol. Biochem. 36: 213-224.

50. Promyou S, Kesta S, Vandoorn W (2008). Hot water treatments delay cold induced banana peel blackening. Postharvest Biol. Technol. 48: 132-138.

51. Purvis AC (1995). Role of alternative oxidase in limiting superoxide production by plant mitochondria. Physiol Plant. 100: 165-170.

52. Purvis AC (2001). Regulation and role of the alternative oxidase in chilling injury of green bell pepper (*Capsicum annuum* L.). Acta Hort. 553: 289-291.

53. Rouet-Mayer MA, Valentova O, Simond-Cote E, Daussant J, Thevenot C (1995). Critical analysis of phospholipid hydrolyzing activities in ripening tomato fruits. Study by spectrofluorimetry and high-performance liquid chromatography. Lipids. 30: 739-46.

54. Raskin I, Turner IM, Melander WR (1989). Regulation of heat production in the inflorescences of an Arum lily by endogenous salicylic acid. PNAS. 86: 2214-2218.

55. Ribas-Carbo M, Aroca R, Gonzàlez-Meler MA, Irigoyen JJ, Sánchez-Diaz M (2000). The electron partitioning between the cytochrome and alternature respiratory pathways during chilling recovery in two cultivars of maize differing chilling sensitivity. Plant Physiol. 122: 199-204.

56. Rui H, Cao S, Shang H, Jin P, Wang K, Zheng Y (2010). Effects of heat treatment on internal browning and membrane fatty acid in loquat fruit in response to chilling stress. J. Sci. Food Agric. 90: 1557-1561.

57. Sung DY, Kaplan F, Lee KJ, Guy CL (2003). Acquired tolerance to temperature extremes. Trends Plant Sci. 8: 179-187.

58. Shewfelt RL, Purvis AC (1995). Toward a comprehensive model for lipid peroxidation in plant tissue. Hortscience. 30: 213-218.

59. Sato Y, Murakami T, Funatsuki H, Matsuba S, Saruyama H, Tanida M (2001). Heat shock-mediated APX gene expression and protection against chilling injury in rice seedlings. J. Exp. Bot. 52: 145-151.

60. Sluse FE, Jarmuszkiewicz W (2000). Activity and functional interaction of alternative oxidase and uncoupling protein in mitochondria from tomato fruit. Braz J. Med Biol Res. 33: 259-68.

61. Sevillano L, Sanchez-Ballesta M, Romojaro F, Flores F (2009). Physiological, hormonal and molecular mechanisms regulating chilling injury in horticultural species. Postharvest technologies applied to reduce its impact. J Sci Food Agric. 89: 555-573.

62. Troncoso-Rojas R, Tiznado-Hernandez ME (2006). Heat shock proteins role in harvested horticultural crops. In: Benkeblia N, Shiomi N, editors. Advances in Postharvest Technologies for Horticultural Crops. India: Res Signpost. pp. 21-39.

63. Thomashow MF (1998). Role of cold-responsive genes in plant freezing tolerance. Plant Physiol.118: 1-7.

64. Valero D, Serrano M (2010). Postharvest Biology and Technology for Preserving Fruit Quality. Boca Raton: CRC-Taylor & Francis.

65. Vanlerberghe GC, Mcintosh L (1994). Mitochondrial electron transport regulation of nuclear gene expression (studies with the alternative oxidase gene of tobacco). Plant Physiol.105: 867-874.

66. Vanlerberghe GC, Day DA, Wiskich JT, Vanlerberghe AE, Mcintosh L (1995). Alternative oxidase activity in tobacco leaf mitochondria (dependence on tricarboxylic acid cycle-mediated redox regulation and pyruvate activation). Plant Physiol. 109: 353-361.

67. Wang X (2001). Plant phospholipases. Annu. Rev. Plant Physiol., Plant Mol. Biol. 52: 211-231.

68. Wagner AM, Krab K (1995). The alternative respiration pathway in plants: role and regulation. Physiol Plant. 95: 318-325.

69. Wonsheree T, Kesta S, Van Doorn WG (2009). The relationship between chilling injury and membrane damage in lemon basil (*Ocimum citriodourum*) leaves. Postharvest Biol. Technol. 51: 91–96.

70. Zheng Y, Fung RW, Wang SY, Wang CY (2008). Transcript levels of antioxidative genes and oxygen radical scavenging enzyme activities in chilled zucchini squash in response to superatmospheric oxygen. Postharvest Biol. Technol. 47: 151-158

71. Zhao D, Shen L, Fan B, Yu M, Zheng Y, Lv S, Sheng J (2009). Ethylene and cold participate in the regulation of LeCBF1 gene expression in postharvest tomato fruits. FEBS Lett. 583: 3329-3334.

Citations

CHAPTER 1

H.S. Hajizadeh and M. Kazemi, 2012, Investigation of Approaches to Preserve Postharvest Quality and Safety in Fresh-cut Fruits and Vegetables, Research Journal of Environmental Sciences, 6: 93-106.

CHAPTER 2

Susan Lurie and Romina Pedreschi, Fundamental Aspects of Postharvest Heat Treatments, doi:10.1038/hortres.2014.30.

CHAPTER 3

Latika Bhatia, Sonia Johri, and Rumana Ahmad, An economic and ecological perspective of ethanol production from renewable agro waste: a review, doi:10.1186/2191-0855-2-65.

CHAPTER 4

Huigang Hu, Xueping Li, Chen Dong, and Weixin Chen, Effects of Wax Treatment on Quality and Postharvest Physiology of Pineapple Fruit in Cold Storage, doi: 10.5897/AJB10.2474.

CHAPTER 5

M. A. Rahman, M. Miaruddin, M. H. H. Khan M. A. T. Masud, and M.M. Begum, Effect of Storage Periods on Postharvest Quality of Pumpkin, ISSN 0258-7122.

CHAPTER 6

Atanda S. A, Pessu P. O, Agoda S., Isong I. U, and Ikotun I, The Concepts and Problems of Post–harvest Food Losses in Perishable Crops, ISSN 1996-0794.

CHAPTER 7

M. A. Rahman, G. M. A. Halim, M. G. F. Chowdhury, M. A. Hossain, and M. M. Rahman, Changes in Physicochemical Attributes of Sweet

Pepper (capsicum annum L.) During Fruit Growth and Development, ISSN 0258-7122.

CHAPTER 8

M. Hailu, T. S. Workneh, and D. Belew, Review on postharvest technology of banana fruit, DOI: 10.5897/AJBX12.020.

CHAPTER 9

Morteza Soleimani Aghdam, Role of Alternative Oxidase in Postharvest Stress of Fruit and Vegetables: Chilling Injury, doi: 10.5897/AJBX11.062.

Index